SpringerBriefs in Earth System Sciences

SpringerBriefs South America and the Southern Hemisphere

W0079779

Series Editors

Gerrit Lohmann
Lawrence A. Mysak
Justus Notholt
Jorge Rabassa
Vikram Unnithan

For further volumes:
http://www.springer.com/series/10032

Claudia P. Tambussi · Federico J. Degrange

South American and Antarctic Continental Cenozoic Birds

Paleobiogeographic Affinities and Disparities

 Springer

Claudia P. Tambussi
División Paleontología Vertebrados
Museo de La Plata
La Plata
Argentina

Federico J. Degrange
CICTERRA/CONICET-UNC
Córdoba
Argentina

ISSN 2191-589X   ISSN 2191-5903   (electronic)
ISBN 978-94-007-5466-9   ISBN 978-94-007-5467-6   (eBook)
DOI 10.1007/978-94-007-5467-6
Springer Dordrecht Heidelberg New York London

Library of Congress Control Number: 2012952140

Printed on acid-free paper

Springer is part of Springer Science+Business Media (www.springer.com)

Preface

"Fossils are, after all, a window to the past and a desperate truth that we will never have an entire clear picture of the ancient times"—Claudia P. Tambussi

This book is the compilation of nearly 30 years of fascination for fossil birds and their evolution. That fascination has maintained my enthusiasm for research on all aspects of birds to the present day. Fortunately, 10 years ago my colleague Federico "Dino" Degrange added his own enthusiasm.

The purpose of this book is to provide a synthesis of the fossil bird record of South America and Antarctica through Cenozoic, considering the geologic forces and climatic–environmental forces that may have shaped its evolution.

The central point is on terrestrial or arboreal birds, but some considerations on aquatic (continental or marine) birds are also made. The temporal focus is on the Paleocene through Pliocene times.

The book begins with a preview (Chaps. 1–4), in which the scope, conceptual, geological, and paleogeographic framework are laid out. The rest is arranged into five main sections. Three chapters compile the information about the main fossil localities chronologically organized. The penultimate Chapter deals with the zoophagous guild, analyzing the record of 13 associations and evaluating the possible dominance of zoophagous over other guilds. The final chapter deals with Bio-Connections of South America and Antarctica with Other Continents and therefore addresses some issues on bird biogeography.

The information came from our own examination of fossils as well as a literature review. The data, however, are not homogeneous, since some temporal gaps without, or with only very scant, information occurs. Some topics inevitably arise in more than one section, but we have tried to keep the repetition to a minimum, and to cross-reference wherever possible.

A second goal of this work is to provide a textbook and source of inspiration for students and novel researchers. We expect that our results have the desired effect. This book is intended to provide a reasonable basis for further research.

During 2011, Dr. Silvio Casadío (CONICET and Universidad Nacional de Río Negro) and Dr. Miguel Griffin (CONICET and Universidad Nacional de La Plata)

organized the symposium *Seaways and landbridges: Southern Hemisphere biogeographic connections through time* in Argentina, which proved to be an interesting space for exchange and discussion on the same topics approached from very different perspectives. This symposium gave us the push to start this work.

Claudia P. Tambussi

Acknowledgments

For providing photographs of fossil specimens, we thank Sandra Chapman. Access to fossil specimens was kindly provided by Jim Holstein, Karsten Lawson, Carl Mehling, Christopher Norris, Dan Brinkman, Luis Chiappe, Kimberly Page Johnson, Marcelo Reguero, Ari Iglesias, Peter Wilf, Alejadro Kramarz, Eduardo Ruigomez, Alejandro Dondas, and Fernando Scaglia. Marcos Cenizo formed the Figs. 2.2, 2.3 and 2.4. We are further obliged to Jorge Rabassa and Sergio Casadío for enabling this book project and Robert Doe and Springer editorial staff for their efforts in the production of this book. This is a contribution to the projects PICT 0365, UNLP N671, PIP-CONICET 0436, and several grants from CONICET and FONCYT from Argentina.

Contents

Abstract

Several advances have been made on the understanding of the biotic and environmental history of South America and Antarctica including the discovery of additional fossil sites coupled with progress from multidisciplinary analyses encompassing tectonic, isotopic, and radiochemical dating and molecular studies in modern forms. This also changed the knowledge about birds. Characters of the South American (SAn) avian fossil record are: (1) presence of taxa with uncertain affinities and absence of Passeriformes during the Paleogene; (2) progressive and accelerated increase of species starting at the Neogene (Miocene); (3) dispersal of important extinct lineages (e.g., Phorusrhacidae, Teratornithidae) to North America after the connection between both Americas; (4) scarce endemic species that are members of clades with major diversification during the Neogene (e.g., Cariamiformes) or that inhabit mainly in the southern hemisphere (e.g., Anhingidae); (5) highly diverse living groups with limited (e.g., Passeriformes) or no (e.g., Apodiformes) fossil record of which stem-groups are registered in Europe; (6) absence of the most extant SAn bird lineages; (7) predominance of the zoophagous birds (>60 %) in all the associations (13) under scrutiny. Changes in diversity of the SAn birds during the Cenozoic could have been the result of the action of different processes (dispersal, vicariance, extirpations, or extinctions) that affect groups in different ways.

Keywords Aves · Cenozoic · South America · Antarctica · Paleobiogeography

Chapter 1
Introduction

Institutional Abbreviations

AMNH	American Museum of Natural History, New York, New York, EEUU
BAR	Museo Asociación Paleontológica Bariloche, Río Negro, Argentina
BMNH	Natural History Museum, London, UK
FM	Field Museum of Natural History, Chicago, Illinois, EEUU
MACN	Museo Argentino de Ciencias Naturales Bernardino Rivadavia, Ciudad Autónoma de Buenos Aires, Buenos Aires, Argentina
MLP	Museo de La Plata, Buenos Aires, Argentina
MPEF-PV	Museo Paleontológico "Egidio Feruglio", Trelew, Chubut, Argentina
MPM-PV	Museo Padre Molina, Río Gallegos, Santa Cruz, Argentina
YPM-PU	Peabody Museum of Natural History, New Haven, Connecticut, EEUU

Modern birds are represented by two big lineages, the Palaeognathae (Tinamiformes + Ratitae) and the Neognathae (Galloanserae + Neoaves) (Mindell and Brown 2005). Fowl and waterfowl (Galloanserae) represent the earliest divergence among neognaths (Fain and Houde 2004) (Fig. 1.1). Both clades sum approximately 10,000 species of which 60 % are Passeriformes (the most diverse clade of terrestrial vertebrates). A comparison between the past and the present reveals a complex and hallmarked evolutionary and biogeographic history which would have begun over 65 million years ago (Tambussi 2011).

The origin of living bird lineages has long been the subject of some controversy. Did living bird lineages originate after the extinction of nonavian dinosaurs at the Cretaceous–Paleogene limit (K-Pg, better known as Cretaceous-Tertiary or K/T boundary)? Or did members of these lineages coexisted with nonavian dinosaurs and survived this great mass extinction event? Whereas the data from biogeography and molecular sequencing argue in favor of the coexistence option, the fossil evidence refutes it, placing the "Big Bang" of avian radiation after the

C. P. Tambussi and F. J. Degrange, *South American and Antarctic Continental Cenozoic Birds*, SpringerBriefs in Earth System Sciences, DOI: 10.1007/978-94-007-5467-6_1, © The Author(s) 2013

Fig. 1.1 Summary of
relationships among main
avian lineages following
Mindell and Brown (2005)

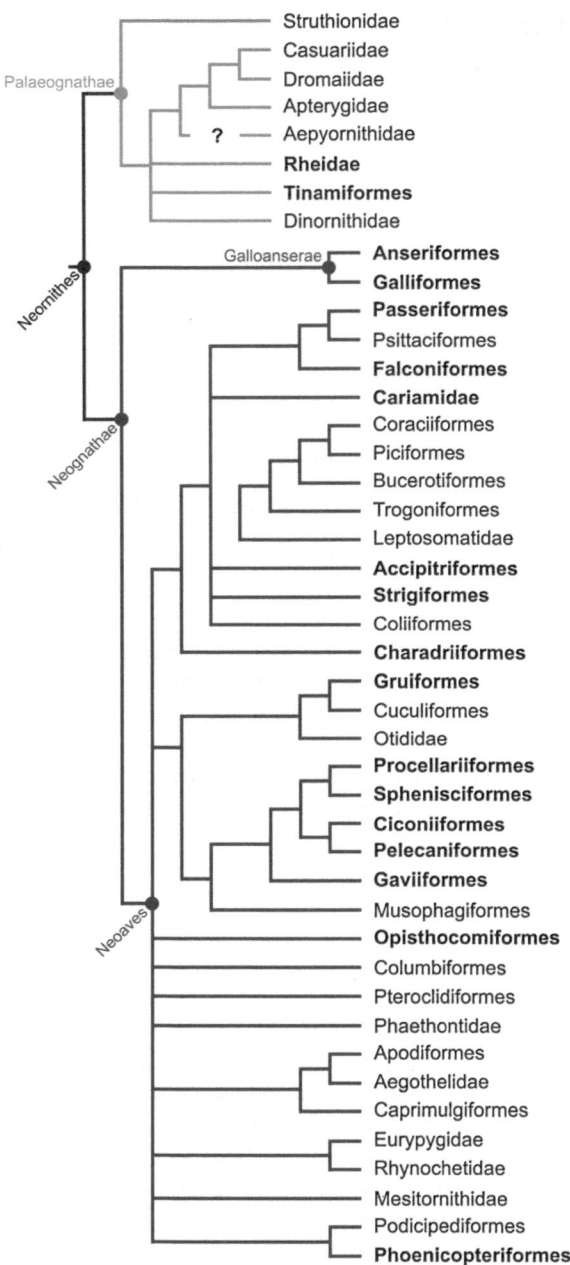

K-Pg boundary (Fig. 1.2). This latter hypothesis is based on two facts: firstly, most
lineages of living birds appear in strata from about 11 to 20 million years ago
following the great extinction event at the end of the Mesozoic, and second there

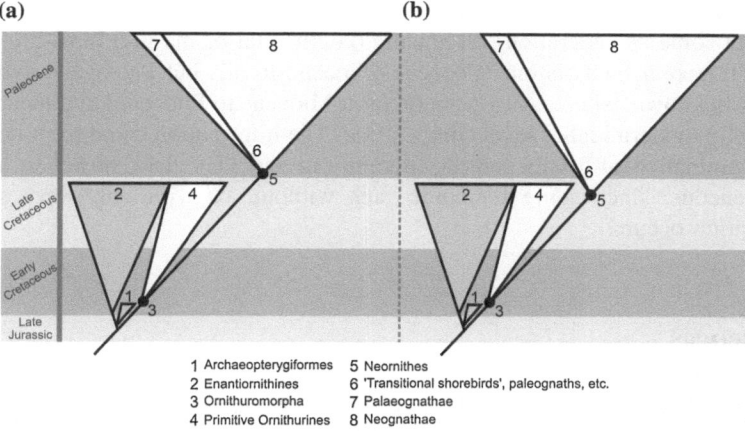

Fig. 1.2 Two modern avian radiation hypotheses. **a** Tertiary radiation hypothesis of Alan Feduccia (1995, 1999). **b** Origin of nearly all modern avian in the Cretaceous, as proposed by Joel Cracraft (1974)

are no Cretaceous fossils remains that can be assigned with certainty to the Neornithes. This scenario was greatly modified some years ago by the finding of a partial skeleton belonging to a new species of Anseriformes named *Vegavis iaai*, about 71 million years old (late Cretaceous, Maastrichtian), from Antarctica, that could be analyzed in a phylogenetic context (Clarke et al. 2005, 2006). From that moment, at least the lineages that include the living screamers, some very primitive geese and the true ducks (Anseriformes), and the close relatives of pheasants and hens (Galliformes) are said to have coexisted with nonavian dinosaurs. This was the first fossil evidence that definitely placed the radiation of modern birds in the Cretaceous. After the K-Pg, the Cenozoic was undoubtedly dominated by mammals and Neornithes birds. Be enough it to say that by the early Oligocene (~35 million years ago), most of the orders of birds that we recognize today had appeared. During recent decades, a great increase in paleornithological information, especially from Miocene through Pleistocene deposits, has become available but knowledge about South American Paleogene birds is almost stagnant. No small bird remains have been recorded so far.

It is quite complicated characterize the effect of environmental changes on bird communities during the South American Cenozoic. In a recent paper, Tambussi (2011) interpreted the paleoenvironmental, paleoecological, and faunal conditions of the Cenozoic using the four most complete bird assemblages recovered from Neogene sediments of Patagonia and the Pampean region. In this work, we summarized the record of land-bird, the paleoenvironmental changes of South America and Antarctica through Cenozoic, emphasizing the relationships between biomes and the geological forces that, through different climatic-environmental factors, have driven its evolution. We increase the area of interest to South

America and Antarctica. The focal point of this analysis is on terrestrial or arboreal birds but some considerations on aquatic (continental or marine) birds are made.

The temporal focus is on Paleocene–Pliocene fossils but Paleogene avifaunas are poorly known, whereas Neogene (at least Miocene to Pliocene) avifauna has an essentially modern higher level composition. The information came from both our own examination of fossils and the literature review. The data, however, are not homogeneous, since some temporal gaps without, or with only very scanty, information occurs.

References

Clarke JA, Tambussi CP, Noriega JI, Erickson GM, Ketcham RA (2005) Definitive fossil evidence for the extant avian radiation in the Cretaceous. Nature 433:305–308

Clarke JA, Tambussi CP, Noriega JI, Erickson GM, Ketcham RA (2006) Corrigendum to definitive fossil evidence for the extant avian radiation in the Cretaceous. Nature 444:780

Cracraft J (1974) Phylogeny and evolution of the ratite birds. Ibis 116:494–521

Fain MG, Houde P (2004) Parallel radiations in the primary clades of birds. Evolution 58:2558–2573

Feduccia A (1995) Explosive evolution in tertiary birds and mammals. Science 267:637–638

Feduccia A (1999) 1,2,3—2,3,4: Accommodating the cladogram. PNAS 96:4740–4742

Mindell DP, Brown JW (2005) Neornithes. Modern birds. http://tolweb.org/Neornithes/15834/ 2005.12.14. Accessed 30 March 2012

Tambussi CP (2011) Paleoenvironmental and faunal inferences based upon the avian fossil record of Patagonia and Pampa: what works and what does not. Biol J Linn Soc 103:458–474

Chapter 2
Paleogeographic Background

The paleogeography of South America is a result of the action of a set of major geological forces such as tectonic, variations in the sea level, sea temperatures, and glaciations (Fig. 2.1), which drove the landscape and climatic evolution of this area (Ortiz Jaureguizar and Cladera 2006). Undoubtedly, these changes go hand in hand with the evolution of the biota. The purpose of this section is to integrate the roles played by these episodes, in shaping the geography and physiognomy of South America. We have focused our attention on the events involved in the formation of deposits with birds that are mentioned in this work.

The most complete Cenozoic South American land-bird fossil record is very largely restricted to Southern South America (SSA—the south of the 15° S area sensu Ortiz Jaureguizar and Cladera 2006), and not just to the earliest Cenozoic but to the latest Paleocene. Deposits with bird remains are distributed geographically across Argentina, Uruguay, Chile, Perú, and Brazil with comparatively few Tertiary land-bird bearing localities outside these countries, e.g., Colombia (Rassmusen and Kay 1992).

The Andes, the longest mountain range in the world, is the outstanding geological feature of South America. It consists of 7,000 km of massive continental rocks all crossing from the north to the south Pacific margin of the continent, which has had deep effects on plant and animal dispersion and evolution in South America. In essence, the Andes represents the tectonic upthrust of rock when the South American plate collides with the Pacific plate. The Southern Andes is the oldest, with significant uplift already prior to the Oligocene. The Central Andes has had most of the uplift in the Miocene or later, whereas the Northern Andes is younger, with its major elevations during the Plio-Pleistocene.

The absence of topographic barriers during Late Cretaceous–Cenozoic times, allowed the Atlantic waters to flood wide areas of the extra-Andean regions (Figs. 2.2 and 2.3). These transgressions occur during the Maastrichtian–Danian,

C. P. Tambussi and F. J. Degrange, *South American and Antarctic Continental Cenozoic Birds*, SpringerBriefs in Earth System Sciences, DOI: 10.1007/978-94-007-5467-6_2, © The Author(s) 2013

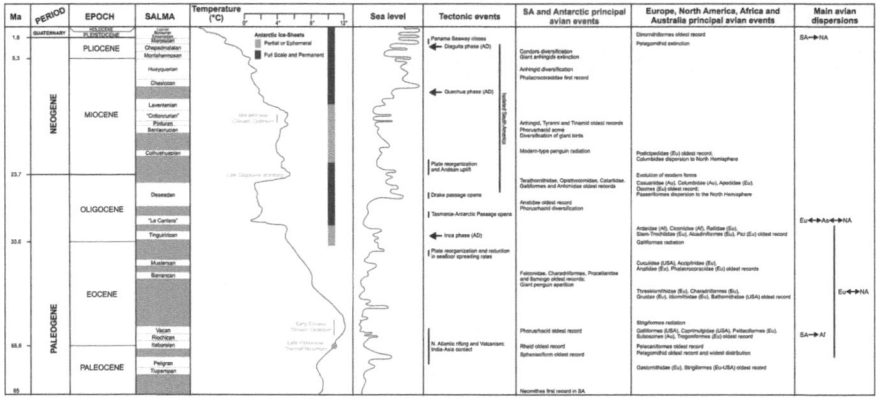

Fig. 2.1 Standard Cenozoic Epochs and some climatic and environmental indicators. Temperature after Zachos et al. (2001), Sea level after Haq et al. (1987), Tectonic events after Pascual et al. (2002) and Zachos et al. (2001), Main avian dispersion after Mayr (2009). *AD* Andean diastrophism; *Af* África; *As* Asia; *Eu* Europe; *NA* North America; *SA* South America

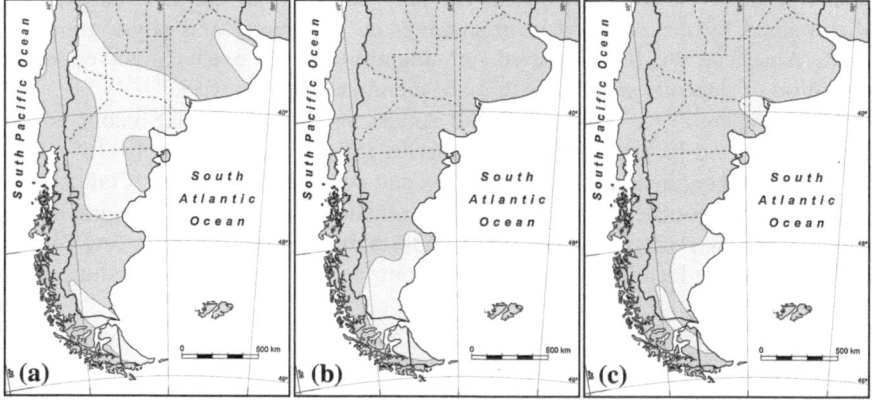

Fig. 2.2 Tentative paleogeography of the marine transgressions in Patagonia between circa 72 and 28 million years ago. **a** Maastrichtian transgression **b** Late Middle Eocene transgression **c** Late Oligocene transgression. Modified from Malumián and Náñez (2011)

Late Middle Eocene, Late Oligocene–Early Miocene, and the Middle Miocene (Malumian and Náñez 2011).

The first transgressive event (Salamancan sea) affected the entire southwest Atlantic basin (Guerstein et al. 2010) that divided southern South America (SSA) into two regions (Fig. 2.2a): the northeastern and the southeastern, respectively (Ortiz Jaureguizar and Cladera 2006). A bridge that linked West Antarctica with SSA still persists revealed by the relationships (sedimentological and faunal) among some Antarctic units (La Meseta, Fossil Hill, and Cross Valley of the James Ross Basin) and the Patagonian units (e.g. Sarmiento, Río Turbio, Cullen, Las

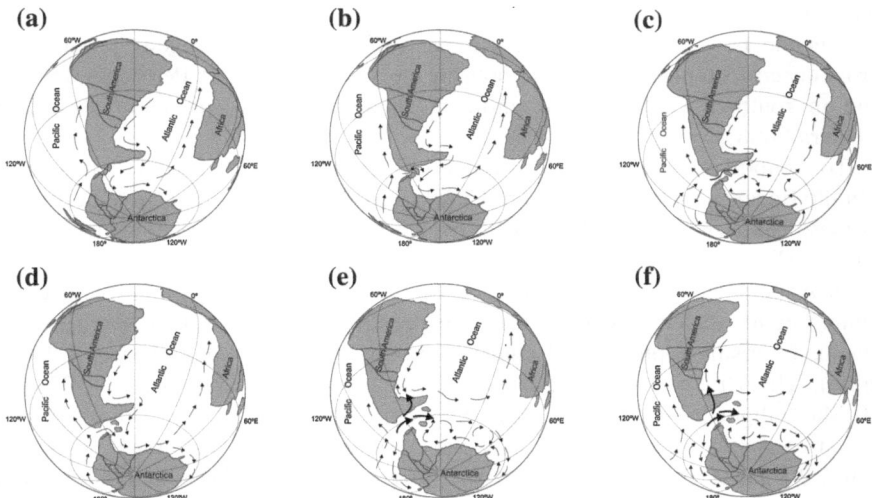

Fig. 2.3 Paleogeographic reconstruction of the major southern landmasses studied in this work, showing the Drake Passage and Atlantic Ocean evolution. **a** 65 Ma, Late Cretaceous–Early Paleocene **b** 51 Ma, Early Eocene Thermal Maximum **c** 34 Ma, Eocene–Oligocene Transition **d** 25 Ma, Late Oligocene **e** 7 Ma, Late Miocene **f** 1.8 Ma, Early Pleistocene. Gray areas indicate landmasses, *black arrows* represents oceanic circulation. Based on Plate Tectonic Reconstructions Online Paleogeographic Mapper (http://www.serg.unicam.it/) and Nullo and Combina (2011)

Flores Formations) (Reguero et al. 2002). During this marine ingression, magmatic and tectonic activity was very low and the epicontinental sea remained in most part of Patagonia until the Danian (Malumian and Náñez 2011).

The areas that were covered by the Salamancan Sea were transformed into flood plains and large lake basins. During the Late Paleocene in Central and Northern Patagonia, large loess plains of pyroclastic sediment developed, while the southernmost tip of Patagonia was covered by water. The Early Paleocene floras were tropical and subtropical forest with mangroves, swampy forest, montagne rain forest, and savanna-sclerophyllous forest, and *Nothophagus* was present but rare (Ortiz Jaureguizar and Cladera 2006; Iglesias et al. 2011). It seems clear that there lived a mix of subtropical with sub-Antarctic elements (Mixed paleoflora, Romero 1986). Toward the end of the Paleocene, there are no records of mangroves.

Deposits of Las Flores (Late Paleocene–Early Eocene) or Salamanca Formations (Early Paleocene) in Patagonia for example, were deposited in these environmental contexts. Almost coeval were deposited farther north, the Itaboraian sediments (Late Paleocene).

The Late Middle Eocene transgression is widely documented in the southern hemisphere but is only recorded in the Austral Basin and offshore of the Colorado Basin in Patagonia. During this time, eustatic sea level was high and seawaters were warm (Zachos et al. 2001), and the occurrence of typical Antarctic foraminifera reveals that temperatures started to fall (Malumián and Náñez 2011). During the Late Eocene, the "Inca Phase" of the Andean orogeny produced a pronounced tectonic deformation both in the Andean basin of Perú and Bolivia, and in southern Chile and Argentina (Tambussi 1989, Ortiz Jaureguizar and Cladera 2006).

To mention one example, the sediments of Laguna del Hunco (Late Paleocene–Middle Eocene) in the northwestern Chubut Province in Argentina were deposited in these environmental contexts, with evidence of high maritime influence on the climate (Wilf et al. 2005).

In Antarctica, during the Early Eocene (50–40 Ma) the separation with Patagonia begins, and the generation of the pre-opening of the Drake Passage that eventually result in physical disconnection between both areas (Scher and Martin 2006). The Drake Passage opened definitively approximately between 32 and 10 Ma (Oligocene–Late Miocene), the Antarctic Circumpolar Current (ACC) began to operate, and concomitantly the Atlantic Ocean temperatures decreased (Figs. 2.1 and 2.3). It has long been recognized that the ACC acts as a barrier, and interrupts the watermass exchange between north to south in the southern oceans (Barnes et al. 2006).

The Late Oligocene and Early Miocene transgressions in Patagonia (Figs. 2.2c and 2.4) produced shallow epicontinental oceans ("Patagonian Sea") with limited extensions, and reveal the existence of cool water current. An updated and detailed description of these transgressions can be found in Bellosi (2010). The Middle Miocene transgression ("Paranean sea") spread to the north of Argentina, covering most of the Chaco-Paraná Basin depression and eastern Patagonia (Hernández et al. 2005) (Fig. 2.4). At Entre Ríos Province (eastern Argentina), the ingressions are represented by the Paraná Formation interpreted as brackish littoral deposits with variable salinity (Aceñolaza and Aceñolaza 2000). At Peninsula Valdes (Chubut Province, Argentina), deposits corresponding to this transgression constitute the Puerto Madryn Formation (Dozo et al. 2010 and the literature cited therein) that consist of cross-bedded sandstones with shells and bioturbated mudstones (Scasso et al. 2001). In the western sectors of SSA, limits of the Miocene transgressions are problematic. For example, the Anta, Río Sali, and San José Formations in northwestern Argentina and Chinches Formation at San Juan Province are unquestionably marine but the relationship between them is much discussed (Hernandez et al. 2005). No bird remains have been recorded so far from the western areas. The continental areas were thus reduced during each transgression, and continental and marine environments coexisted.

Basaltic volcanism in Patagonia began during the Maastrichtian, continued during the Cenozoic, and affected wide areas of Patagonia. The highest volcanic activity occurs from the Paleocene to the Eocene (Panza and Franchi 2002), followed by a Late Oligocene basaltic lava event (29–25 Ma), and a Late Miocene to Early Pliocene one (16–5 Ma) that affected central to southern Patagonia. The lava produced from this volcanic activity added to pyroclastic materials were deposited as part of the continental sequences (Nullo and Combina 2011) that characterized most of the Patagonian paleontological sites. For example, Chichinales and Collun

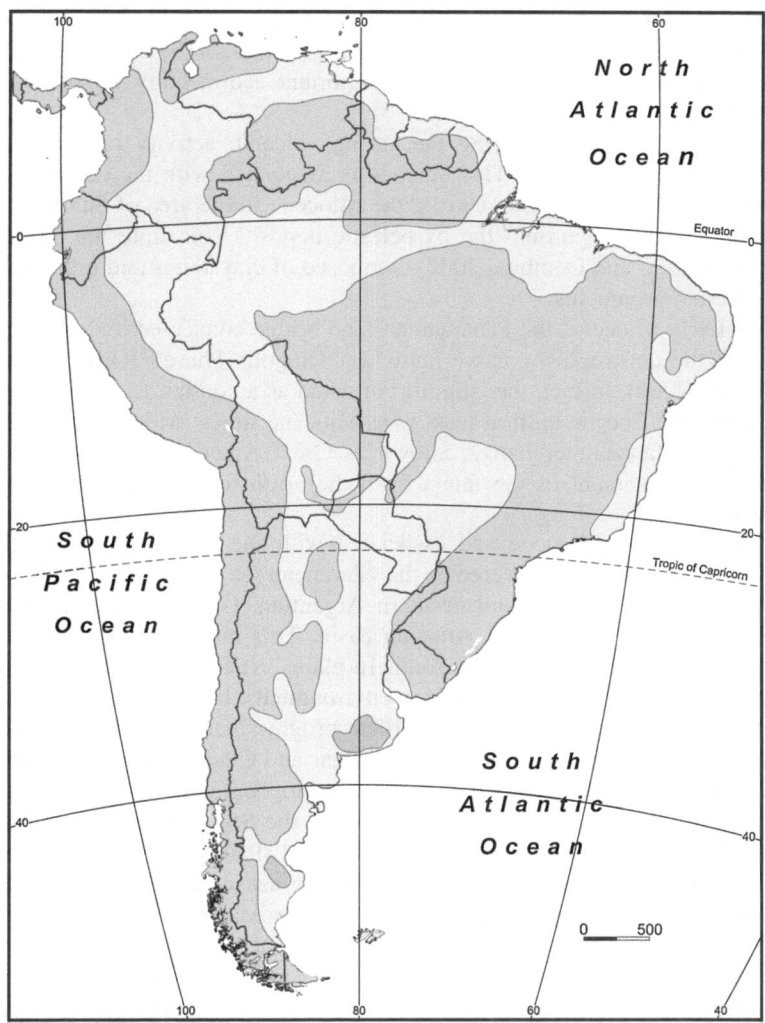

Fig. 2.4 Tentative paleogeography of the marine Middle–Late Miocene Paranense transgression between 15 and 13 Ma. Modified from Donato et al. (2003)

Curá Formations (Northern Patagonia), Sarmiento, Pinturas, and Santa Cruz Formation (central and southern Patagonia) are continental deposits from the Miocene to Pliocene with high frequency of volcanic elements.

Across all central–southern Patagonia, various continental sequences from Eocene to Early Miocene (35–19 Ma) are visible. The characteristic mammal remains from these sequences ("Toba mammals") are important to understand the evolution of the entire South American continent (Flynn and Swisher 1995; Pascual et al. 2002). In Patagonia, the "Musters Formation", Early Oligocene in age and the coetaneous Abanico Formation in Chile (containing the Tinguiririca fauna) had wide continental distribution with grasslands. The Colhue Huapi Formation (Chubut, Argentina) is another important sedimentary sequence of the same age.

Basaltic lava produced during the strong volcanic activity of the Miocene, covered wide areas of SSA. This episode is associated with the collision of the Chile ridge with the continent. During the Miocene in the area of Santa Cruz and Tierra del Fuego (Argentina) the pyroclastic deposits constitute the Santa Cruz Formation (Nullo and Combina 2002) composed of claystones and tuffs typical of continental environments.

In the Early Miocene, the Panamanian land bridge connected both Americas as a result of the diastrophism as we know as "Diaguita Phase" (Ortiz Jaureguizar and Cladera 2006). In fact, the isthmus consisted as a continuous chain above sea level from Late Eocene until at least Late Miocene times (Montes et al. 2012).

During the fauna interchange, known as Great American Biotic Interchange (GABI), the movement of the fauna from the north to the south was dominant (Woodburne 2010).

During the Late Miocene and associated with the "Quechua Phase" of the Andean orogeny, the areas covered by the "Paranean sea" were succeeded by plains reaching Patagonia, central and northern Argentina, Uruguay, Bolivia, southern Perú, Venezuela and the upper Amazon basin. This marks the beginning of the episode known as the "Age of the southern plains" (Pascual and Bondesio 1982) characterized by high frequency of open environments. In the Pliocene this was the heyday of this event. The Andean cordillera progressively acts as a barrier of the moisture-laden Pacific winds (Ortiz Jaureguizar and Cladera 2006) and the differentiation between Subantarctic and Patagonian biogeographic subregions began.

From the Early Paleocene to the Pleistocene the SSA environments showed a climate change from warm, wet, and nonseasonal (Paleocene to Eocene) to cold and dry (Middle Eocene to Early Oligocene) to seasonal climate (Middle–Late Miocene) (Ortiz Jaureguizar and Cladera 2006; Barreda and Bellosi 2003; Barreda and Palazzesi 2007) (Fig. 2.1).

In a sequence that includes subtropical forest, savanna woodland, park-savanna, and savanna grassland, the Paleocene tropical forests were replaced by the steppes that now strongly characterize the extra-Andean Patagonia (Barreda and Palazzesi 2007) (Fig. 2.5). Iglesias et al. (2007) estimate the annual mean temperatures between 12 and 15 °C and palaeoprecipitations of 1,100 mm during the Paleogene. A continuous global warming is observed during the Paleogene with two pinnacles: the Late Paleocene (LPTM $\sim$ 56 Ma) and the Early Eocene (EECO $\sim$ 52 Ma) optima (Fig. 2.1). The global average palaeotemperatures were 10 °C higher than those currently recorded for South America (Zachos et al. 2001) with a smaller difference in temperature from the equator to the poles. The increase in the global temperatures ended in the Eocene in relation to the early opening of the Drake and Tasmania Passages between Antarctica/South America and Antarctica/Australia respectively, which allow the circum-Antarctic circulation, causing global temperature decreases that strongly affected SSA (Figs. 2.1 and 2.3).

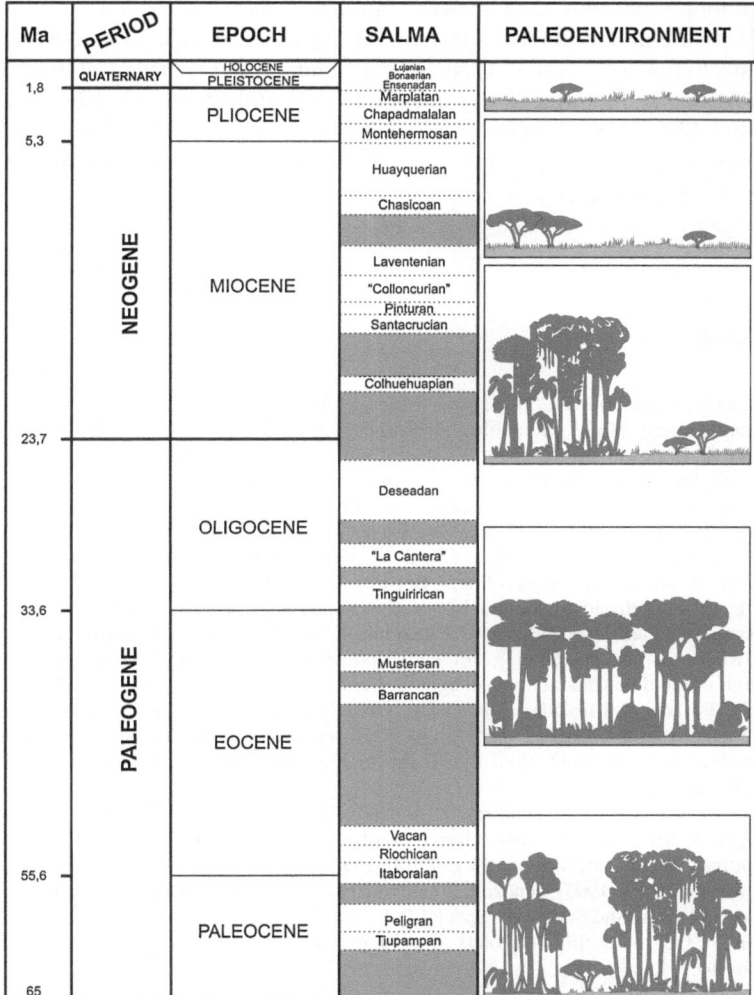

Fig. 2.5 Palaeocene to recent timescale following Gradstein et al. (2004) including timescale for Cenozoic mammalian faunas of South America (SALMA) showing vegetation-type physiognomies under increasingly drier and/or more markedly seasonal climates. Paleoenvironments were taken from Barreda and Palazzesi (2007)

A third event of temperature increase takes place toward the end of Oligocene (Zachos et al. 2001); it is called Late Oligocene warming (LOW, Fig. 2.1).

Since the Oligocene, all biogeographical regions previously recognized migrated to lower latitudes (Iglesias et al. 2011). For the Miocene, the first expansion of herbaceous shrub is noticed and began the development of extreme aridity and seasonality in eastern Patagonia (Fig. 2.5).

For a long time, from the Cretaceous to the Neogene, West Antarctica and SA (Magallanic Region) remained attached forming an independent continent isolated by oceans to the north and east (Nullo and Combina 2011). This isolation in turn, gave a particular footprint to the fauna and flora (Olivero et al. 1990, Marenssi et al. 1994, Shen 1995, Reguero et al. 1998, 2002). From the EECO to the Eocene–Oligocene transition (~34 Ma) temperature was a descent into an icehouse climate. During this decrease ice began to reappear at the poles, and Antarctic ice sheet began to rapidly expand (Fig. 2.1).

References

Aceñolaza FG, Aceñolaza G (2000) Trazas fósiles del Terciario marino de Entre Ríos (Formación Paraná, Mioceno Medio), República Argentina. Bol Acad Nac Cienc 64:209–233

Barnes DKA, Hodgson DA, Convey P, Allen CS, Clarke A (2006) Incursion and excursion of Antarctic biota: past, present and future. Global Ecol Biogeogr 15:121–142

Barreda V, Bellosi E (2003) Ecosistemas terrestres del Mioceno temprano de la Patagonia central: primeros avances. Rev Mus Arg Cienc Nat 5:125–134

Barreda V, Palazzesi L (2007) Patagonian vegetation turnovers during the paleogene-early neogene: origin of arid adapted floras. Bot Rev 73:31–50

Bellosi E (2010) Physical stratigraphy of the Sarmiento formation (middle Eocene-lower Miocene) at Gran Barranca, central Patagonia. In: Madden RH, Carlini AA, Vucetich MG, Kay RF (eds) The paleontology of Gran Barranca: evolution and environmental change through the middle Cenozoic of Patagonia. Cambridge University Press, New York

Donato M, Posadas P, Miranda-Esquivel DR, Ortiz Jaureguizar E, Cladera G (2003) Historical biogeography of the Andean region: evidence from Listroderina (Coleoptera: Curculionidae: Rhytirrhinini) in the context of the South American geobiotic scenario. Biol J Linn Soc 80:339–352

Dozo T, Bouza P, Monti A, Palazzesi L, Barreda V, Massaferro G, Scasso R, Tambussi CP (2010) Late Miocene continental biota in northeastern Patagonia (Península Valdés, Chubut, Argentina). Palaeogeog Palaeoecol 297:100–109

Flynn J, Swisher C III (1995) Cenozoic South American land mammal ages: correlation to global geochronologies. Soc Sed Geol, Special Pub 54:317–333

Gradstein F, Ogg J, Smith A (2004) A geologic time scale. Cambridge University Press, Cambridge

Guerstein GR, Guler MV, Brinkhuis H, Warnaar J (2010) Mid Cenozoic palaeoclimatic and palaeoceanographic trends in the Southwest Atlantic basins, a dinoflagellate view. The paleontology of Gran Barranca. In: Madden RH, Carlini AA, Vucetich MG, Kay RF (eds) The paleontology of Gran Barranca: evolution and environmental change through the middle Cenozoic of Patagonia. Cambridge University Press, Cambridge

Haq BU, Hardenbol J, Vail PR (1987) Chronology of fluctuating sea levels since the triassic (250 million years ago to present). Science 235:1156–1167

Hernández RM, Jordan TE, Dalenz Farjat A, Echavarría L, Idleman BD, Reynolds JH (2005) Age, distribution, tectonics, and eustatic controls of the Paranense and Caribbean marine transgressions in southern Bolivia and Argentina. J South Am Earth Sci 19:495–512

Iglesias A, Wilf P, Johnson K, Zamuner A, Cuneo NR, Matheos S (2007) A paleocene lowland macroflora from Patagonia reveals significantly greater richness than North American analogs. Geology 35:947–950

Iglesias A, Artabe AE, Morel EM (2011) The evolution of Patagonian climate and vegetation, from the Mesozoic to the present. Biol J Linn Soc 103:409–422

Malumián N, Nañez C (2011) The late cretaceous-cenozoic transgressions in Patagonia and the Fuegian Andes: foraminifera, paleoecology and paleogeography. Biol J Linn Soc 103:269–288

Marenssi SA, Reguero MA, Santillana SA, Vizcaíno SF (1994) Eocene land mammals from Seymour Island, Antarctic palaeobiogeographical implications. Antarctic Sci 6:3–15

Mayr G (2009) Paleogene fossil birds. Springer, Berlin

Montes C, Bayona G, Cardona A, Buchs DM, Silva CA, Morón S, Hoyos N, Ramírez DA, Jaramillo CA, Valencia V (2012) Arc-continent collision and orocline formation: closing of the Central American seaway. J Geophys Res. doi:10.1029/2011JB008959

Nullo F, Combina A (2002) Sedimentitas terciarias continentales. In: Haller MJ (ed) Geología y Recursos naturales de Santa Cruz. Buenos Aires

Nullo F, Combina A (2011) Patagonian continental deposits (Cretaceous-Tertiary). Biol J Linn Soc 103:289–304

Olivero EB, Gasparini Z, Rinaldi CA, Scasso R (1990) First record of dinosaurs in Antarctica (Upper Cretaceous, James Ross Island): palaeogeographical implications. In: Thomson MRA, Crame JA, Thomson JW (eds) Geological Evolution of Antarctica. Cambridge University Press, Cambridge.

Ortiz Jaureguizar E, Cladera GA (2006) Paleoenvironmental evolution of southern South America during the cenozoic. J Arid Environ 66:498–532

Panza JL, Franchi MR (2002) Magmatismo basáltico Cenozoico extrandino. In: Haller MJ (ed) Geología y Recursos Naturales de Santa Cruz. XV Congreso Geológico Argentino

Pascual R, Bondesio P (1982) Un roedor Cardiatheriinae (Hydrochoerydae) de la Edad Huayqueriense (Mioceno tardío) de La Pampa. Sumario de los ambientes terrestres en la Argentina durante el Mioceno. Ameginiana 29:19–35

Pascual R, Carlini AA, Bond M, goin FJ (2002) Mamíferos cenozoicos. In: Haller MJ (ed) Geologia y Recursos Naturales de Santa Cruz. Asociación Geológica Argentina, Buenos Aires

Rasmussen DT, Kay RF (1992) A Miocene anhinga from Colombia, and comments on the zoogeographic relationships of South America's Tertiary avifauna. Sci Series 36:225–230

Reguero MA, Vizcaíno SF, Goin FJ, Marenssi SA, Santillana SN (1998) Eocene high-latitude terrestrial vertebrates from Antarctica as biogeographic evidence. Asoc Pal Arg 5:185–198

Reguero MA, Marenssi SA, Santillana SN (2002) Antarctic Peninsula and Patagonia Paleogene terrestrial environments: biotic and biogeographic relationships. Palaeogeog Palaeoecol 2776:1–22

Romero E (1986) Paleogene phytogeography and climatology of South America. Ann Miss Bot Gard 73:449–461

Scasso R, McArthur JM, del Río C, Martínez S, Thirlwall MF (2001) 87Sr/86Sr Late Miocene age of fossil molluscs in the "Entrerriense" of the Valdés Peninsula (Chubut, Argentina). J South Am Earth Sci 14:319–329

Shen Y (1995) Subdivision and correlation of Cretaceous to Paleogene Volcano-sedimentary sequence from Fildes Peninsula, King George Island, Antarctica. In: Shen Y (ed) Stratigraphy and Paleontology of Fildes Peninsula, King George Island, Antarctica. State Antarctic Committee Monograph 3, Beijing Science Press.

Scher HD, Martin EE (2006) Timing and climatic consequences of the opening of Drake Passage. Science 312:428–430

Tambussi CP (1989) Las aves del Plioceno tardío-Pleistoceno temprano de la Provincia de Buenos Aires. Dissertation, Universidad Nacional de La Plata

Wilf P, Johnson KR, Cúneo NR, Smith ME, Singer BS, Gandolfo MA (2005) Eocene plant diversity at Laguna del Hunco and Río Pichileufú. Patagonia, Argentina Am Nat 165:634–650

Woodburne MO (2010) The Great American biotic interchange: dispersals, tectonics, climate, sea level and holding pens. J Mamm Evol 17:245–264

Zachos J, Shackleton NJ, Revenaugh JS, Pälike H, Flower BP (2001) Climate response to orbital forcing across the oligocene-miocene. Science 292:274–278

Chapter 3
Geological Settings of the Major Fossil Localities in South America and Antarctica

Aves remains have been recovered from several continental Cenozoic localities in SSA, mainly in the Pampas and Patagonia in Argentina, Uruguay, and Brazil. We have selected 13 localities because they have the most complete fossil bird associations (Fig. 3.1). We use here the term "association" when specimens and taxa belong to the same sedimentary formation although not necessarily from the same locality (but nearby; e.g., Killik Aike Norte, Estancia La Costa, and Puesto Estancia La Costa localities are all considered Santacrucian in age).

The São José de Itaboraí Basin or Itaboraí Basin is a rich fossiliferous locality of Southeastern Brazil about 60 km from the city of Rio de Janeiro. The sediments are represented by limestone rocks that were vertically incised by fissures where the fossils were found (Bergqvist et al. 2011). The taphocoenosis contains mammals, reptiles, birds, amphibians, plants, gastropods, palynomorphs, and ostracods (Bergqvist et al. 2008, 2011). Based on the rich fossil mammal assemblage, it was proposed Itaboraian South American Mammal Age (SALMA) by Marshall (1985) belonging to the late Paleocene (approximately 60 Ma Pascual and Ortiz Jaureguizar (2007), but see Gelfo et al. 2009 provisionally regard its age as early Eocene, 2 or 4 million years younger than hitherto supposed). Four species of extinguished birds were described from Itaboraí: two big-sized terrestrial birds, *Diogenornis fragilis* and *Paleopsilopterus itaboraiensis* (see Alvarenga 1983, 1985a) and two small landbirds *Eutreptodactylus itaboraiensis* and *Itaboravis elaphrocnemoides* (Baird and Vickers- Rich 1997; Mayr et al. 2011). Nowadays, the Itaboraí Basin is completely covered with water impeding any collecting activity (Kellner and Campos 1999).

During the end of the nineteenth century, Ameghino (1865–1936) between 1887 and 1902 collected fossil avian remains of the "Deseadan" (Oligocene) in different localities of southern Patagonia, but the situation of these localities was imprecise, maybe because of the absence of accurate cartography of the area (Carlini et al. 2010). "Deseadan" sediment surface in large areas of the lower

C. P. Tambussi and F. J. Degrange, *South American and Antarctic Continental Cenozoic Birds*, SpringerBriefs in Earth System Sciences, DOI: 10.1007/978-94-007-5467-6_3,

Fig. 3.1 Important South American and Antarctic avian fossil localities studied in this work

course of Rio Deseado at Santa Cruz Province (Argentina), and the Deseado Formation defined by Loomis (1914) corresponds to what is currently defined as Sarmiento Formation. Most of the birds of Oligocene age that are known were described by (Ameghino 1895, 1899). In all cases, specimens are very fragmentary with difficult and controversial assignment. As far as we know, there has been no new records of this age the 1840s (e.g., *Andrewsornis abbotti*).

The Tremembé Formation in the Taubaté basin in eastern São Paulo State, Brazil, dated as Late Oligocene to Early Miocene based on its fossil content, is a lacustrine sedimentary unit, essentially composed of dark shales (Kellner and Campos 1999). Tremembé Formation is divided into two section formations, Tremembé and Caçapava, the former being highly fossiliferous. Actinopterygian fishes, amphibians, turtles, snakes, caimans, mammals, and at least nine species of birds recognized until now (Alvarenga 1982, 1985b, 1988, 1990, 1993, 1995) were recovered from the Tremembé Formation.

According to the biostratigraphic data provided in the literature, the Santa Cruz Formation is of late Early Miocene (Vizcaíno et al. 2010 and the literature cited therein). This formation was originally recognized for its exposures at the southern end of the Atlantic coast (Tauber 1997a, b) with a radiometric dating yield of 16.53 Ma (Fleagle et al. 1995). The sediments of the Santa Cruz Formation were deposited in estuarine, fluvial, and eolian environments during the final regression of the Superpatagonian Sea (Bellosi 2010).

In western central Patagonia, in the Río Pinturas valley, terrestrial sediments of late Early to Middle Miocene age constitutes the Pinturas Formation. This Formation corresponds to a continental sequence with mainly tuffaceous sediments. It is divided into three sequences by two erosive intraformational discordances. Lower and middle sequences have Colhuehuapan and Santacrucian faunas together with some exclusive species. However, the upper sequence presents only typical santacrucian species. Some authors have correlated these deposits with those from the Santa Cruz Formation and it has even been included in the same (Pascual and Odreman Rivas 1971; Marshall 1976a, b); meanwhile, other authors state that it is only possible to correlate in part these deposits with those of the Santa Cruz Formation (Kramarz and Bellosi 2005). Nevertheless, Ré et al. (2010) have demonstrated through radiometric datation that Pinturas Formation is slightly older than the Santa Cruz Formation (Ameghino 1906; Frenguelli 1931; Barrio et al. 1984; Fleagle et al. 1995). There are several paleoambiental contradictions in regard to the lower and middle sequences of this formation. Meanwhile, palynological and faunal data point out the presence of humid forests (Chiappe 1991); sedimentological, paleopedological, and ichnological data seem to indicate subhumid to semiarid conditions. These conditions are also inferred for the upper sequence.

A considerable important bird association (Cenizo et al. 2011) is known from several fossiliferous localities of central and northern La Pampa province (central Argentina), in sediments assigned to the Late Miocene Cerro Azul Formation (Salinas Grandes de Hidalgo, Laguna Guatraché, Quehué, Bajo Giuliani, El Guanaco, Caleufú, and Buenos Aires) (Folguera and Zárate 2009). The analysis of the sediments indicates the existence of lacustrine deposits at the base, which overlies eolic levels (Linares et al. 1980; Goin et al. 2000). The Cerro Azul Formation is characterized by its lithological homogeneity, mostly formed by silts, sandy silts, and very fine silty sands, pinkish to reddish brown, and evidence of pedogenesis (Folguera and Zárate 2009). Geochronology of this unit falls between 10 Ma and 5.8 to 5.7 million years (Visconti et al. 2010).

The Ituzaingó Formation outcrops along the Paraná River in eastern Argentina and bears a vertebrate assemblage with strong affinities with the Acre fauna (Brazil) and the correlative Kiyu Formation (Uruguay) (Cione et al. 2000; Latrubesse et al. 2007; Perea et al. 1994). The sandy–clayey and conglomeratic sediments of the Tertiary Ituzaingó Formation have been apparently deposited in meandering river and marshes. The underlying marine and/or estuarine Paraná Formation is regarded as Tortonian in age (Late Miocene) according to Aceñolaza and Aceñolaza (2000). The exact age of the Ituzaingó Formation has been largely discussed. As proposed by Cione et al. (2000), it is assumed that the fauna of the "Conglomerado osífero" (the lower level of the Formation) is assigned to the Late Miocene, pending a comprehensive study of correlation with other fossiliferous units of the Neogene of South America (Candela and Noriega 2004). Above this formation, Pleistocene continental sediments of Ensenadan and Lujanian age were deposited (Iriondo 1980; Cione and Tonni 1995, 1996; Herbst 2000).

The Santa María Group (Middle Miocene–Late Pliocene) outcrops in the northwest of Argentina and it has been divided into six formations, from base to top: San José, Las Arcas, Chiquimil, Andalhualá, Corral Quemado, and Yasyamayo (Bossi and Palma 1982). Except for the last two, all these formations are fossil bearers (invertebrates, vertebrates, and plants). The lower portion of the Santa María Group would have direct connection with the ingression of the "Paranense" sea. It is probable that the fluvial plains had received successive contributions of marine water that flooded these plains, generating huge brackish water bodies. Nevertheless, no exclusive marine fossils have been recorded, although some microfossils may indicate hypersalinity in some sectors, which is explained by the sporadic contributions of marine water (Herbst et al. 2000). Andalhualá Formation (Late Miocene–Early Pliocene) is basically constituted by upward-coarsening sandstones, with abundant conglomerates and a few pelitic levels, and diverse tuffaceous layers (Anzótegui et al. 2007; Bossi and Muruaga 2009). It comprises the lower and middle Araucanian of Frenguelli (1930) and the Araucanian horizon of Riggs and Patterson (1939), the Araucanense Formation of Butler et al. (1984) and part of the El Cajón Formation (Turner 1962). The paleoenvironment could have consisted in meandering rivers of lower sinuosity (Bossi and Palma 1982; Bossi et al. 1998) that produced sedimentary deposition. It is the thickest formation and the one that possesses the highest diversity in the Group (Marshall and Patterson 1981). It has contributed a huge amount of fossil plants (Anzótegui et al. 2007) but it is richest in fossil vertebrates, such as birds, reptiles, and especially mammals (Marshall and Patterson 1981; Nasif 1998; Bossi et al. 1999; Herbst et al. 2000; Herrera and Ortiz 2005; Agnolín 2006, 2009). This vertebrate fauna has been interpreted as belonging to the Huayquerian SALMA (9–6.68 Ma) (Pascual and Odreman Rivas 1971). Nevertheless, the temporary limits for the Andalhualá Formation are 7 and 3.54 Ma, according to Herrera and Ortiz (2005). This fact leads to think that the Santa María Valley (where Andalhualá Formation outcrops), would have a relictual Huayquerian fauna during the development of the Montehermosan age (such as the fauna of Corral Quemado Formation). This older fauna could have persisted due to the isolation produced by the existence of geographic or climatic barriers (Herrera and Ortiz 2005). Beds overlying the Andalhualá Formation possibly are Chapadmalalan and Montehermosan in age (see Cione and Tonni 1996).

The coastal deposits exposed between Mar del Plata and Miramar together with the classical Monte Hermoso locality represent the most complete Cenozoic stratigraphic sequence across the Pampean Region in Argentina. Concerning the Pliocene, a well-known bird association (Tambussi 2011) was found in litho-stratigraphic units included in the Chapadmalal Formation, Chapadmalalan stage (Late Pliocene, ca 4.0–3.0 Ma, early Late Pliocene sensu Candela and Rasia 2012) exposed in the southern coast of the Buenos Aires Province. As other continental late Cenozoic deposits in central Argentina, Chapadmalal Formation is charac-terized by a relatively homogeneous sedimentary sequence composed of fine volcanoclastic sediments and massive brownish silts (Deschamps et al. 2011). These sequences are characterized by its colorimetry and its sedimentological homogeneity makes it difficult to recognize the boundaries between the different formations. Kraglievich (1952) defined seven formational units, Chapadmalal, Barranca de Los Lobos, Vorohué, Miramar, Arroyo Seco, Santa Isabel, and Lobería Formations. In a first and excellent summary of these units, Zárate (1989) recognized two lithostratigraphic units, Pampean and above, the eolian Lobería Formation. Also he identified five alloformations (ALF): Playa San Carlos, San Andrés, Punta Martinez de Hoz, Punta San Andrés, and Arroyo Lobería. The first two correspond to the Chapadmalal Formation, the third to Barranca de Los Lobos, the fourth to Vorohué, San Andrés, Miramar, and Arroyo Seco, and Arroyo Lobería ALF to the Arroyo Lobería Formation (Zanchetta 1995). The Monte Hermoso Formation (Farola Monte Hermoso at Buenos Aires Province, Early Pliocene) is very important because it is the type locality of biostratigraphic units of the Pliocene South American mammal deposits. Recently, Deschamps et al. (2011) noted the dissimilarity between upper levels of the Monte Hermoso Formation (Unit II, Lower Chapadmalalan) and those of the Chapadmalal Formation (Upper Chapadmalalan).

To conclude this section, we need to mention two marine deposits that host important continental bird remains: La Meseta (Seymour Island, Antartica pen-insula) and Puerto Madryn (Patagonia) Formations.

The Puerto Madryn Formation at Península Valdés (northeastern Patagonia) is a Late Miocene clastic sedimentary sequence well-known for its abundant content of marine invertebrates and well-preserved vertebrate remains including teleost fishes, penguins, and marine mammals. A new vertebrate faunal assemblage was recently discovered from two new localities in the southwestern coast of Península Valdés (Rincón Chico and La Pastosa paleontological sites) that belong to the uppermost portion of the Puerto Madryn Formation (Dozo et al. 2010 and references cited therein). According to Scasso and del Río (1987), this portion was accumulated in a shallow shelf environment. Stratigraphically, both deposits correspond to the "Rionegrense" (uppermost levels of the Puerto Madryn Formation or Río Negro Formation, according to different authors). Sediments of La Pastosa site consists of interstratified sandstones and shales forming hetherolithic facies, with intercalated coquinas composed by disarticulated oyster and pectinid valves in a sandy matrix, and conglomerates. Vertebrate remains come from the later that consists in oligomictic, intraformational conglomerate with matrix-supported texture and mud intraclasts (Dozo et al. 2010). Sandstones with shale intercalations or shale lenses dominate the profile of Rincón Chico site which also includes some conglomerate lenses and coquina levels similar to those in La Pastosa.

The La Meseta Formation, in James Ross Basin on the East side of the Antarctic Peninsula, is a rich fossiliferous marine deposit composed of sandstones, mudstones, and conglomerates deposited during the Eocene in deltaic, estuarine, and shallow marine settings (Marenssi et al. 1998a, b; Tambussi et al. 2006). Six units are distinguished from the base to the top (Marenssi et al. 1998b): Valle de Las Focas, Acantilados, Campamento, Cucullaea I, Cucullaea II, and Submeseta Allomembers. They are grouped into three facies and altogether represent a major transgressive cycle. Dingle and Lavelle (1998) reported a ^{87}Sr/^{86}Sr derived age of 34.2 Ma (late Late Eocene) for the topmost part of La Meseta Formation, whereas Dutton et al. (2002) reported ages of 36.13, 34.96, and 34.69 Ma (late Late Eocene) for different levels within Submeseta Allomember. These strata document the highest morphological and taxonomical diversity of penguins in the world that lived sympatrically (Tambussi et al. 2006). High concentration of penguin bones is especially located between two shell banks (Myrcha et al. 1990): a lower one bearing the gastropod *Turritella* and a higher one bearing bivalve *Modiolus* and the brachiopod *Lingula*. This zone, having five penguin species exclusively recorded in this interval, was determined as *Anthropornis nordenskjoeldi* Biozone by Tambussi et al. (2006). It is easily distinguishable by the common occurrence of penguin bones and the phosphatic brachiopod *Lingula*.

References

Aceñolaza FG, Aceñolaza G (2000) Trazas fósiles del Terciario marino de Entre Ríos (Formación Paraná, Mioceno Medio), República Argentina. Bol Acad Nac Cienc 64:209–233

Agnolín FL (2006) Notas sobre el registro de Accipitridae (Aves, Accipitriformes) fósiles argentinos. Stud Geol Salmant 42:67–80

Agnolín FL (2009) Una nueva especie del genero Megapaloelodus (Aves: Phoenicopteridae: Palaelodinae) del Mioceno superior del noroeste de Argentina. Rev Mus Arg Cienc Nat 11:23–32

Alvarenga HMF (1982) Uma gigantesca ave fóssil do Cenozóico brasileiro: Physornis brasiliensis sp. n. An Acad bras Ciênc 54:697–712

Alvarenga HMF (1983) Uma ave ratite do Paleoceno brasileiro: bacia calcária de Itaboraí, estado do Rio de Janeiro, Brasil. Bol Mus Nac do Rio de Jan, Geol 41:1–47

Alvarenga HMF (1985a) Um novo Psilopteridae (Aves: Gruiformes) dos sedimentos Terciários de Itaboraí, Rio de Janeiro, Brasil. Anais do Cong Bras de Pal 8. Série Geol 27:17–20

Alvarenga HMF (1985b) Notes on the Cathartidae (Aves) and description of a new genus from the Brazilian Cenozoic. An Acad bras Ciênc 57:349–357

Alvarenga HMF (1988) Ave fóssil (Gruiformes: Rallidae) dos folhelhos da Bacia de Taubaté, Estado de São Paulo, Brasil. An Acad bras Ciênc 60:321–332

Alvarenga HMF (1990) Flamingos fosseis da bacia de Taubaté, estado de São Paulo. Descricao de nova especie: An Acad bras Ciênc, Brasil 62:335–345

Alvarenga HMF (1993) Paraphysornis novo gênero para Physornis brasiliensis Alvarenga, 1982 (Aves: phorusrhacidae). An Acad bras Ciênc 65:403–406

Alvarenga HMF (1995) A large and probably flightless anhinga from the Miocene of Chile. Cou For Senck 181:149–161

Ameghino F (1895) Sobre las aves fósiles de Patagonia. Bol Inst Geog Arg 15:501–602

Ameghino F (1899) Sinopsis geológico-paleontológica de la Argentina. Segundo Censo Rep Arg 1:112–255

Ameghino F (1906) Les formations sédimentaires du Crétacé supérieur et du Tertiaire de Patagonie. An Mus Nac Buenos Aires 15:v

Anzótegui LM, Garralla SS, Herbst R (2007) Fabaceae de la Formación El Morterito, (Mioceno Superior) del valle del Cajón, provincia de Catamarca, Argentina. Ameghiniana 44:183–196

Baird RF, Vickers-Rich P (1997) Eutreptodactylus itaboraiensis gen. et. sp. nov., an early cuckoo (Aves: Cuculiformes) from the Late Paleocene of Brazil. Alcheringa 21:123–127

Barrio RE, Scillato-Yane G, Bown M (1984) La Formación Santa Cruz en el borde occidental del macizo del Deseado (Provincia de Santa Cruz) y su contenido paleontológico. Actas Cong Geol Arg 1:539–556

Bellosi E (2010) Physical stratigraphy of the Sarmiento Formation (middle Eocene-lower Miocene) at Gran Barranca, central Patagonia. In: Madden RH, Carlini AA, Vucetich MG, Kay RF (eds) The paleontology of Gran Barranca: evolution and environmental change through the middle Cenozoic of Patagonia. Cambridge University Press, New York

Bergqvist LP, Mansur K, Rodrigues MA, Rodrigues-Francisco BH, Perez RAR, Beltrão MCMC (2008) Itaboraí Basin, state of Rio de Janeiro-the cradle of mammals in Brazil. In: Schobbenhaus C, Souza CRG, Fernandes ACS, Berbert-Born M, Queiroz ET (eds) Winge M. Sítios Geológicos e Paleontológicos do Brasil, Brasil

Bergqvist LP, Batista de Almeidae E, De Araújo JUNIOR H (2011) Tafonomia da assembleia fossilífera de mamíferos da "fenda1968", Bacia De São José de Itaboraí, Estado do Rio de Janeiro, Brasil. Rev Bras Paleontol 14:75–86

Bossi GE, Palma RM (1982) Reconsideración de la estratigrafía del Valle de Santa María, Provincia de Catamarca, Argentina. V Cong Latin Geol 1:155–172

Bossi GE, Gavriloff I, Esteban G (1998) Terciario. Estratigrafía, bioestratigrafía y paleogeografía. In: Gianfrancisco M, Puchulu M, Durango de Cabrera J, Aceñolaza GE (eds) Geología de Tucumán, Tucumán

Bossi G, Muruaga CM, Georgieff S (1999) El Neógeno del faldeo occidental del cerro Pampa, Catamarca. XIV Cong Geol Arg, Actas I:483–486

Bossi GE, Muruaga CM (2009) Estratigrafía e inversión tectónica del 'rift' neógeno en el Campo del Arenal, Catamarca, NO Argentina. And Geol 36:311–341

Butler RF, Marshall LG, Drake RE, Curtis GH (1984) Magnetic polarity stratigraphy and ^{40}K–^{40}Ar dating of Late Miocene and early pliocene continental deposits, Catamarca Province, NW Argentina. J Geol 92:623–636

Candela AM, Noriega JI (2004) Los coipos (Rodentia, Caviomorpha, Myocastoridae) del "Mesopotamiense" (Mioceno tardío; Formación Ituzaingó) de la provincia de Entre Ríos, Argentina. In: Aceñolaza FG (ed.) Temas de la Biodiversidad del Litoral Fluvial Argentino, serie INSUGEO 12

Candela AM, Rasia LL (2012) Tooth morphology of Echimyidae (Rodentia, Caviomorpha): homology assessments, fossils, and evolution. Zool J Linn Soc 164:451–480

Carlini AA, Ciancio MR, Scillato-Yané GJ (2010) Middle Eoene-early Miocene Dasypodidae (Xenarthra) of South America: faunal succession at Gran Barranca-biostratigraphy and paleoecology. In: Madden RH, Carlini AA, Vucetich MG, Kay RF (eds) The paleontology of Gran Barranca: evolution and environmental change through the middle Cenozoic of Patagonia. Cambridge University Press, New York

Cenizo MM, Tambussi CP, Montalvo CI (2011) Late Miocene continental birds from the Cerro Azul formation in the Pampean region (central-southern Argentina). Alcheringa, pp 1–22

Chiappe LM (1991) Fossil birds from the Miocene Pinturas Formation of southern Argentina. J Vert Pal 17:21–22A

Cione AL, Tonni EP (1995) Bioestratigrafía y cronología del Cenozoico superior de la región pampeana. In: Alberdi MT, Leone G, Tonni EP (eds) Evolución biológica y climática de la región pampeana durante los últimos cinco millones de años: un ensayo de correlación con el Mediterráneo occidental. Monografías del Museo Nacional de Ciencias Naturales, Madrid

Cione AL, Tonni EP (1996) Reassessment of the Pliocene-Pleistocene continental time scale of southern America. Correlation of Chapadmalalan with Bolivia sections. J South Am Earth Sci 9:221–236

Cione AL, Azpelicueta M, Bond M, Carlini AA, Casciotta JR, Cozzuol MA, de la Fuente M, Gasparini Z, Goin FJ, Noriega J, Scillato-Yané GJ, Soilbelzon L, Tonni EP, Verzi D, Vucetich MG (2000) Miocene vertebrates from Entre Ríos Province, eastern Argentina. In: Aceñolaza FG, Herbst R (eds) El Neógeno de Argentina. Insugeo, Serie Correlación Geológica 14

Deschamps CM, Vucetich MG, Verzi DH, Olivares AI (2011) Biostratigraphy and correlation of the Monte Hermoso formation (early Pliocene, Argentina): the evidence from caviomorph rodents. J South Am Earth Sci 35:1–9

Dingle RV, Lavelle M (1998) Antarctic Peninsular cryosphere: early Oligocene (c. 30 Ma) initiation and a revised glacial chronology. J Geol Soc 155:433–437

Dozo T, Bouza P, Monti A, Palazzesi L, Barreda V, Massaferro G, Scasso R, Tambussi CP (2010) Late Miocene continental biota in northeastern Patagonia (Península Valdés, Chubut, Argentina). Palaeogeog Palaeoecol 297:100–109

Dutton AL, Lohmann KC, Zinsmeister WJ (2002) Stable isotope and minor element proxies for Eocene climate of Seymour Island. Paleocean, Antarctica. doi:10.1029/2000PA000593

Fleagle JG, Bown TM, Swisher C, Buckley G (1995) Age of the Pinturas and Santa Cruz Formations. Actas VI Cong Arg Pal Bioestrat 1:129–135

Folguera A, Zárate M (2009) La sedimentación Neógena continental en el sector extrandino de Argentina Central. Rev Asoc Geol Arg 64:692–712

Frenguelli J (1930) Las guayquerías de San Carlos en la provincia de Mendoza. Pub Dep Ext Univ 9:1–54

Frenguelli J (1931) Nomenclatura estratigráfica patagónica. An Soc Cient Sta Fe 3:1–115

Gelfo JN, Goin FJ, Woodburne MO, de Muizon C (2009) Biochronological relationships of the earliest South American Paleogene mammalian faunas. Paleontology 52:251–269

Goin FJ, Montalvo CI, Visconti G (2000) Los Marsupiales (Mammalia) del Mioceno Superior de la Formación Cerro Azul (provincia de La Pampa, Argentina). Estud Geol 56:101–126

Herbst R (2000) La Formación Ituzaingó (Plioceno). Estratigráfica y distribución. Corr Geol 14:181–190

Herbst R, Anzótegui LM, Esteban G, Mautino LR, Morton S, Nasif N (2000) Síntesis paleontológica del Mioceno de los valles Calchaquíes, Noroeste Argentino. Corr Geol 14:263–288

Herrera CM, Ortiz PE (2005) Nuevos registros de mamíferos para la Formación Andalhualá (Mioceno tardío-Plioceno), provincia de Tucumán. Ameghiniana 42:33R

Iriondo MH (1980) El Cuaternario de Entre Ríos. Rev Asoc Cienc Nat Lit 11:125–141

Kraglievich L (1952) El perfil geológico de Chapadmalal y Miramar, Provincia de Buenos Aires. Rev Mus Mun Cienc Nat Trad Mar del Plata 1:8–37

Kellner AWA, Campos DA (1999) Vertebrate paleontology in Brazil: a review. Episodes 22: 238–251

Latrubesse EM, Silva SAF, Cozzuol MA, Absy ML (2007) Late Miocene continental sedimentation in southwestern Amazonia and its regional significance: biotic and geological evidence. J South Am Earth Sci 23:61–80

Linares E, Llambías EJ, Latorre CO (1980) Geología de la provincia de La Pampa, República Argentina y geocronología de sus rocas metamórficas y eruptivas. Rev Asoc Geol Arg 35:87–146

Loomis F (1914) The Deseado Formation of Patagonia. The Rumford press, Concord

Kramarz AG, Bellosi ES (2005) Hystricognath rodents from the Pinturas Formation, Early-Middle Miocene of Patagonia, biostratigraphic and paleoenvironmental implications. J South Am Earth Sci 18:199–212

Marenssi SA, Santillana SV, Rinaldi CA (1998a) Stratigraphy of La Meseta Formation (Eocene), Marambio Island, Antarctica. In: Casadío S (ed) Paleógeno de America del sur y de la Península Antártica. Revista de la Asociación Paleontológica Argentina 5

Marenssi SA, Santillana SV, Rinaldi CA (1998b) Paleoambientes sedimentarios de la Aloformación La Meseta (Eoceno), isla Marambio (Seymour), Antártida. Inst Ant Arg, cont 464

Marshall LG (1976a) On the affinities of Pichipilus osborni Ameghino 1890 (Marsupialia, Caenolestinae) from the Santa Cruz Beds of Southern Patagonia, Argentina. Ameghiniana 13:56–64

Marshall LG (1976b) Fossil localities for Santacruzian (early Miocene) mammals, Santa Cruz provincie. Southern Patagonia Argentina. J Pal 50:1129–1142

Marshall LG (1985) Geochronology and Land-mammal biochronology of the transamerican faunal interchange. In: Stehli D, Webb SD (eds) The great american interchange. Plenum Press, New York

Marshall LG, Patterson B (1981) Geology and geochronology of the mammal-bearing Tertiary of the valle de Santa María and río Corral Quemado Catamarca province, Argentina. Fieldiana Geol 9:1–80

Mayr G, Alvarenga HMF, Clarke J (2011) An Elaphrocnemus-like landbird and other avian remains from the late Paleocene of Brazil. Acta Palaeontol Pol 56:679–684

Myrcha A, Tatur A, Del Valle R (1990) A new species of fossil penguin from Seymour Island, West Antarctica. Alcheringa 14:195–205

Nasif N (1998) Nuevo material de Eumysopinae (Echimyidae, Rodentia) del Terciario Tardío (Formación Andalhualá), Valle de Santa María, Provincia de Catamarca, Argentina. Ameghiniana 35:3–6

Pascual R, Odreman Rivas O (1971) Evolución de las comunidades de de los vertebrados del Terciario argentino. Los aspectos paleozoogeográficos y paleoclimáticos relacionados. Ameghiniana 8:372–412

Pascual R, Ortiz-Jaureguizar E (2007) The Gondwanan and South American Episodes: two major and unrelated moments in the history of the South American mammals. J Mammal Evol 14:75–137

Perea D, Ubilla M, Martínez S, Piñeiro G, Verde M (1994) Mamíferos neógenos del Uruguay: la edad Mamífero Huayqueriense en el "Mesopotamiense". Acta Geol Leopol 17:375–389

Ré GH, Bellosi ES, Heizler M, Vilas JF, Madden RH, Carlini AA, Kay RF, Vucetich MG (2010) A geochronology for the Sarmiento Formation at Gran Barranca. In: Madden RH, Carlini AA, Vucetich MG, Kay RF (eds) The paleontology of Gran Barranca: evolutional and environmental change through the middle Cenozoic of Patagonia. Cambridge University Press, New York

Riggs ES, Patterson B (1939) Stratigraphy of Late-Miocene and Pliocene deposits of the province of Catamarca (Argentina) with notes on the faunae. Physis 14:143–162

Scasso R, del Río C (1987) Ambiente de sedimentación, estratigrafía y proveniencia de la secuencia marina del Terciario superior de la región de península de Valdés. Rev Asoc Geol Arg 42:291–321

Tambussi CP (2011) Paleoenvironmental and faunal inferences based upon the avian fossil record of Patagonia and Pampa: what works and what does not. Biol J Linn Soc 103:458–474

Tambussi CP, Acosta Hospitaleche C, Reguero MA, Marenssi SA (2006) Late Eocene penguins from West Antarctica: systematics and biostratigraphy. In: Francis JE, Pirrie D, Crame JA (eds) Cretaceous-Tertiary high-latitude palaeoenvironments, James Ross Basin, Antarctica. Geological Soc London, London

Tauber AL (1997a) Bioestratigrafía de la Formación Santa Cruz (Mioceno inferior) en el extremo sudeste de la Patagonia. Ameghiniana 34:413–426

Tauber AL (1997b) Paleoecología de la Formación Santa Cruz (Mioceno inferior) en el extremo sudeste de la Patagonia. Ameghiniana 34:517–529

Turner JCM (1962) Estratigrafía del tramo medio de la sierra de Velasco y región al oeste (La Rioja). Bol Acad Nac Cienc Córd 43:37

Visconti G, Melchor R, Montalvo C, Umazano A, De Elorriaga E (2010) Análisis litoestratigráfico de la Formación Cerro Azul (Mioceno Superior) en la provincia de La Pampa. Rev Asoc Geol Argent 67:257–265

Vizcaíno SF, Bargo MS, Kay RF, Fariña RA, Di Giacomo M, Perry JMG, Prevosti FJ, Toledo N, Cassini GH, Fernicola JC (2010) A baseline paleoecological study for the Santa Cruz Formation (late-early Miocene) at the Atlantic coast of Patagonia, Argentina. Palaeogeog Palaeoecol 292:507–519

Zanchetta G (1995) Estado actual de la geología y estratigrafía de los depósitos Plio-Pleistocenos de la Región Bonaerense. In: Alberdi MT, Leone G, Tonni EP (eds) Evolución biológica y climática de la Región Pampeana durante los últimos cinco millones de años. Un ensayo de correlación con el Mediterráneo Occidental, Monografías del Museo Nacional de Ciencias Naturales, Madrid

Zárate MA (1989) Estratigrafía y geología del Cenozoico Tardío en los acantilados marinos comprendidos entre Playa San Carlos y A° Chapadmalal, Partido de General Pueyrredón. Dissertation, Universidad Nacional de La Plata, Prov. De Buenos Aires

Chapter 4
The Nature of the Fossil Record of Birds

No science is based on complete information, and paleornithology is not the exception. There are several reasons for these gaps in the knowledge. The avian fossil record is certainly incomplete; bird remains constitute a biased data set, the discovery of fossils is not haphazard, and those found reveal only limited information, usually about skeletal morphology only. But they are unique reliable sources of information that document features and changes in past life.

For centuries, biologists have stated that avian bones are hollow and delicate; even school texts note that the skeleton of birds is delicate and light to compensate the high energy cost of flying. This notion of lightness of the bird skeleton had already been pointed out by Galileo (Galileo 1863 in Dumont 2010) and ever since that time, the idea that this lightness is correlated with fragility and consequently with low skeletal resistance was established. In the field of paleontology, it is a well-known fact that fossil bird remains are substantially less frequent than those of other vertebrate groups, and the most common plausible explanation for this fact has been the fragility of bird bones. This notion is currently tottering.

The structure of bird bones consists of a strong and thin external cortex supported by internal struts and pneumatic bone (Fig. 4.1), contrasting with the typical mammalian structure that consists of a thick cortical layer supported by spongy bone. Despite these differences, the bone tissue that makes up the bones of birds is dense, even more than that of mammals, and this high density of the cortical bone is correlated with fewer pores per surface unit compared with other types of bony tissues of higher mineral concentration (Bonser 1995). The novel information provided very recently by Elizabeth Dumont (2010) from her study of the density of the skull, humerus or upper arm bone, and femur or thigh bone, radically alters the idea we used to have about the qualities of the bird skeleton and compels its redefinition. Despite their delicate appearance, bird skeletons are not lightweight relative to total body or soft tissue mass. In fact, the weight of the skeleton of a 50 g songbird (Passeriformes) is similar to that of a rodent of equal weight. In fact, the average/mean bone density calculated from the skull, femur

C. P. Tambussi and F. J. Degrange, *South American and Antarctic Continental Cenozoic Birds*,   25
SpringerBriefs in Earth System Sciences, DOI: 10.1007/978-94-007-5467-6_4,
© The Author(s) 2013

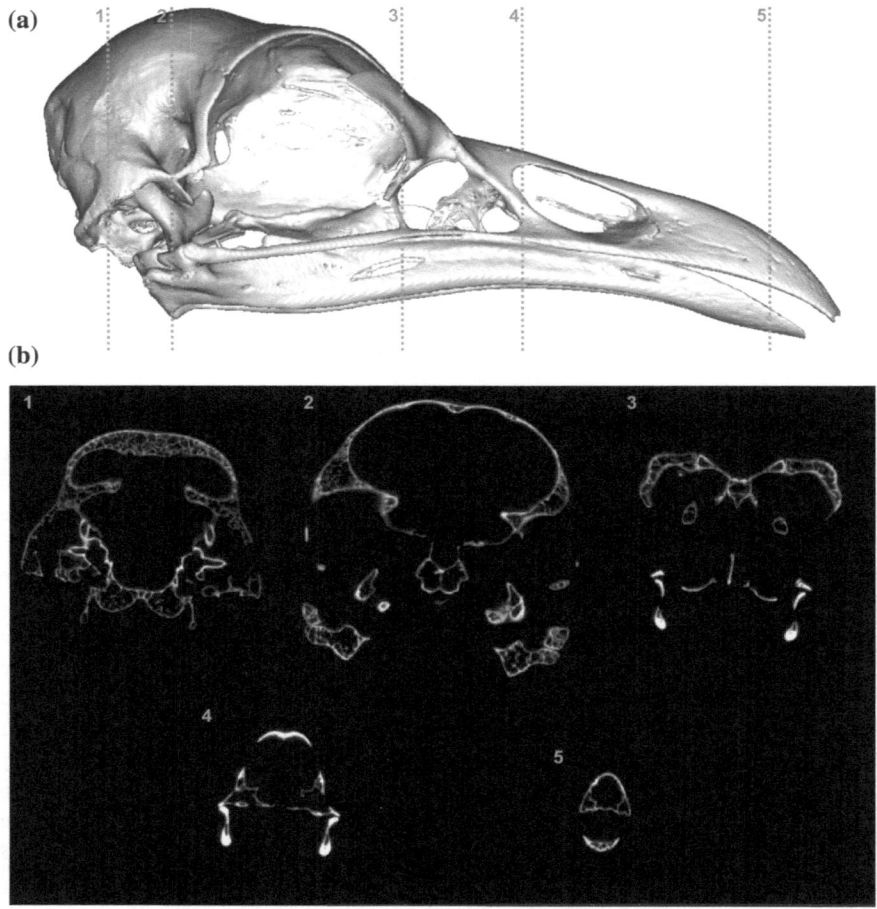

Fig. 4.1 Structure of bird bones: **a** 3d model of the skull of *Cariama cristata* in lateral view, *vertical dotted lines* 1–5 indicate transverse slices shown in (**b**)

and humerus, and weight to assess the relative contribution of each of these elements to the total weight of the skeleton, is surprisingly higher in birds than in mammals. Thus, the bird skeleton appears to be delicate and fragile but relatively stronger and more rigid due to its bone density (mass per volume unit). This is not exactly true in the case of the femur (the bone most closely associated to support the body mass), whose density in birds is somewhat lower than in bats and small rodents (Dumont 2010). In any case, bird bones are hollow but dense, and as this density increases, so does their resistance and rigidity. This new perspective contrasts with the one that explains the low frequency of birds in the fossil record compared to other vertebrates, as a result of the supposed fragility of their skeleton, and suggests that this phenomenon probably involves other factors.

The probability that a carcass is totally or partially preserved as a fossil in deep time is closely related to the processes that take place immediately after death, the manner of the latter, and the climate conditions at the time of decease (Gardner and Walker 2009). Bioerosion is one of the most powerful postmortem destructive processes (Davis 1997) in terms of the speed and depth of its action (Trueman and Martill 2002) and it is infrequently observed in fossilized bones.

Contemporary taphonomists are deeply concerned with the study of the decay, disarticulation, damage, and transport of skeletons, and their work is essential to explain adequately the processes intervening in the composition of the fossil record (Bickart 1984; Cione et al. 2011; Cruz 2003, 2005, 2006, 2007, 2008). A priori, it would seem that the processes of skeleton disarticulation and disaggregation are very fast (Cruz 2003) and strongly favor the generation of the most characteristic items of the fossil record of Cenozoic birds: isolated bones.

One highly interesting aspect is the fact that animals with robust or pachyostotic bones present a more complete fossil record because of their higher preservation potential. At the same time, these bones are more susceptible to suffer spatial mixture due to transport processes and time-averaging, because they can remain unburied for long periods. This idea agrees with Kowalewski's (1997) reciprocal taphonomic model. It can be exemplified with penguins (Sphenisciformes), whose fossil record in Patagonia or Antarctica is doubtlessly the most complete among birds (Acosta Hospitaleche et al. 2008; Acosta Hospitaleche and Reguero 2010; Jadwiszczak 2010; Myrcha et al. 2002; Tambussi et al. 2006; Tonni 1980). Additionally, penguins are gregarious and bigger populations are sure to produce more fossils.

As we pointed out, the South American Cenozoic bird fossil record consists largely of isolated bones, although some nearly complete skeletons have been recovered from certain localities (Tambussi 2011). The remains of 'non-penguins', although less likely to be preserved, provide a record with better spatial–temporal resolution (see Section—Major Fossil Localities and Deposits). In the Paleogene the record of continental birds is scarce and fragmentary. A growing number of taxa are recorded since the Miocene, in fact most of the bird orders that we recognize today had appeared during that period (Tambussi et al. 1993; Tambussi 2011). In all cases, the specimens are macroscopic possibly resulted of the style of sampling mode in the field, reflecting only partially the bird diversity.

References

Acosta Hospitaleche C, Reguero MA (2010) First articulated skeleton of Palaeeudyptes gunnari from the late Eocene of Isla Marambio (Seymour Island), Antarctica. Antarctic Sc 22: 289–298
Acosta Hospitaleche C, Castro L, Tambussi CP, Scasso R (2008) *Palaeospheniscus patagonicus* (Aves, Spheniscidae): new discoveries from the Early Miocene of Argentina and its paleoenvironmental significance. J Pal 82:565–575
Bonser RH (1995) Longitudinal variation in mechanical competence of bone along the avian humerus. J Exp Zool 198:209-212.

Bickart KJ (1984) A field experiment in avian taphonomy. J Vert Pal 4:525–535

Cione AL, Cozzuol MA, Dozo MT, Acosta Hospitaleche (2011) Marine vertebrate assemblages in the Southwest Atlantic during the Miocene. Biol J Linn Soc 103:423-440

Cruz I (2003) Paisajes tafonómicos de restos de Aves en el sur de Patagonia continental. Dissertation, Universidad de Buenos Aires, Aportes para la interpretación de conjuntos avifaunísticos en registros arqueológicos del Holoceno

Cruz I (2005) La representación de partes esqueléticas de aves. Patrones naturales einterpretación arqueológica. Int J Archaeozool 14:69–81

Cruz I (2006) Los restos de pingüinos (Spheniscidae) de los sitios de Cabo Blanco (Santa Cruz, Patagonia Argentina). Análisis tafonómico y perspectivas arqueológicas. Int Antropol 7:15–26

Cruz I (2007) Avian taphonomy: observations at two Magellanic penguin (*Spheniscus magellanicus*) breeding colonies and their implications for the fossil record. J Archaeol Sci 34:1252–1261

Cruz I (2008) Avian and mammalian bone taphonomy in Southern Continental Patagonia, a comparative approach. Quat Int 180:30–37

Davis PG (1997) The bioerosion of bird bones. Int J Osteoarchaeol 7:388–401

Dumont E (2010) Bone density and the lightweight skeletons of birds. Proc Biol Sci 277:2193-8

Gardner E, Walker S (2009) Climate matters: comparing avian bone taphonomy in warm temperate vs subtropical environments. Portland GSA Annual Meeting, Portland

Jadwiszczak P (2010) New data on the appendicular skeleton and diversity of Eocene Antarctic penguins. In: Nowakowski D (ed) Morphology and systematics of fossil vertebrates. DN Publisher, Poland

Kowalewski M (1997) The Reciprocal Taphonomic Model. Lethaia 30:86-88

Myrcha A, Jadwiszczak P, Tambussi CP, Noriega JI, Gaździcki A, Tatur A, del Valle RA (2002) Taxonomic revision of Eocene Antarctic penguins based on tarsometatarsal morphology. Polish Polar Research 23:5-46

Tambussi CP (2011) Paleoenvironmental and faunal inferences based upon the avian fossil record of Patagonia and Pampa: what works and what does not. Biol J Linn Soc 103:458–474

Tambussi CP, Acosta Hospitaleche C, Reguero MA, Marenssi SA (2006) Late Eocene penguins from West Antarctica: systematics and biostratigraphy. In: Francis JE, Pirrie D, Crame JA (eds) Cretaceous-Tertiary high-latitude palaeoenvironments, James Ross Basin, Antarctica. Geological Society of London, London

Tambussi CP, Noriega JI, Tonni EP (1993) Late Cenozoic birds of Buenos Aires Province: an attempt to document quantitative faunal changes. Palaeogeog Palaeoecol 101:117–129

Tonni EP (1980) The present state of knowledge of the Cenozoic birds of Argentina. Contrib Sci 330:104–114

Trueman CN, Martill DM (2002) The long term survival of bone: the role of bioerosion. Archaeometry 44:371–382

Chapter 5
The Paleogene Birds of South America

For modern lineages of birds, few fossils have been found that predate the Cretaceous–Palaeogene (K–Pg) boundary, 65 million years ago. However, molecular studies using fossil calibrations have shown that many of these lineages existed at that time (Smith et al. 2011 and literature cited therein). Based on this evidences, two entrenched ideas circulate on the evolution of modern birds. One supports that after the Cretaceous–Tertiary transition witnessed a major ordinal diversification within extant birds; the other sustain that the diversification occur deep within the Mesozoic (Fig. 1.2). Countless publications show alternatively one idea or the other (Brown et al. 2008; Clarke et al. 2005; Fountaine et al. 2005; Ericson et al. 2006; Van Tuinen et al. 2006), and many attempts to reconcile them have failed.

In any case, the fossil record of South American Paleogene birds does not help much in this controversy. Fossils of alleged neornithine birds are sparse and fragmentary, inconclusive, and their phylogenetic assignment is usually controversial (Tambussi 2011).

5.1 Paleocene

The earliest record of Neornithes for South America comes from Palacio de los Loros locality (Chubut, Argentina) in sediments of the Salamanca Formation for the Danian-Selandiano limit (~ 61.7 Ma) (Degrange et al. 2006). Specimens consist of two downy feathers preserved on their part and counterpart (Fig. 5.1 a–d) found in association with a large amount of plant material, mainly angiosperms (Iglesias et al. 2007, 2011). These feathers have symmetrical blade and raquis relatively short but because the base is absent more precise identification is not possible (Dove com. pers). Assuming that all Cenozoic birds are Neornithes we presuppose that the feathers belong to this clade.

C. P. Tambussi and F. J. Degrange, *South American and Antarctic Continental Cenozoic Birds*, SpringerBriefs in Earth System Sciences, DOI: 10.1007/978-94-007-5467-6_5,
© The Author(s) 2013

Fig. 5.1 Paleocene feathers from: **a–d** Salamanca Formation for the Danian-Selandiano limit (~61.7 Ma) in Palacio Los Loros locality and (**e**) Late Paleocene of Maíz Gordo Formation (Santa Barbara Subgroup) in the La Mendieta locality, both at Argentina. **a** Downy feathers preserved on their part and counterpart MPEF-FL-1A, **b** Pennaceous distal portion viewed at SEM (Scanning Electron Microscope), **c** Detail of the barbules, **d** MPEF-FL-1B, **e** Feather with oblique barbs of 18 mm long, barbules and hooks preserved

Fossil feathers are also known from Late Paleocene of Maíz Gordo Formation (Santa Barbara Subgroup) in the La Mendieta locality, Jujuy, Argentina (Petrulevicius and Tambuscci 1995). The feather is fragmented, the calamus is absent, the raquis is 25 long with oblique barbs of 18 mm long, barbules, and hooks are preserved. Again, we presuppose that the feathers belong to Neornithes (Fig. 5.1e).

Putative Rheiformes were reported from the middle Paleocene of Las Flores locality, middle section of Río Chico Formation, Chubut, Argentina (Tambussi 1989). Features of the preserved pedal phalanges indicate that the specimen belongs to the living morphotype (Tambussi 1995), different from that of the Itaboraian rheids (see below).

Fossil birds from the Late Paleocene fissure filling in São José de Itaboraí in Brazil are few but very significant. In fact, Itaboraian birds are the oldest South American fossil Neornithes association represented by skeletal remains. This fauna is characterized by taxa with no or little flight capabilities such as Rheiformes and Cariamiformes and other birds of uncertain affinities among which are the only small terrestrial birds found across South America during this period (Mayr et al. 2011a).

Diogenornis fragilis represents one of the earliest records of Rheiformes of South America (Tambussi 1995). Several remains of this species have been collected (Alvarenga 1983, Carvalho deTaranto et al. 2011), among which there are fragments of the forelimb, several vertebrae, tarsometatarsus, and tibiotarsus. From the size of their bones it can be assumed that *Diogenornis* was of small size (~80–90 cm), truly terrestrial with apparent inability to fly, with a beak different from living rheas but similar to the Galliformes (Alvarenga 1993). The proportions of the bones and particularly of the humerus—less reduced than in other Rheiformes—indicate that *Diogenornis* belongs to a different morphotype that other Rheiforms (Tambussi 1995). Although it was originally positioned in the family Opisthodactylidae, Mayr (2009) assigns it to Rheidae and highlights the similarity between the tarsometatarsus of *Diogenornis* with the European Palaeognathae Palaeotididae and Remiornithidae. More recently, Alvarenga (2010) highlighted the possible relation of *Diogenornis* with the Casuariidae. In the same manner as all other taxa from the same locality, *Diogenornis* is exclusive of Itaboraí Basin.

Itaboarian Cariamiformes are *Itaboravis elaphrocnemoides* and *Paleopsilopterus itaboraiensis* named by Mayr et al. (2011) and Alvarenga (1985a) respectively.

Traditionally, a number of terrestrial and wading bird families that did not seem to belong to any other order were classified together as Gruiformes. Such are the cases of Gruidae, Rallidae, Heliornithidae, Psophiidae, and Cariamidae among many other families. Now is accepted that living Cariamidae (seriemas) should be placed in its own Order, Cariamiformes. There are two living species of Cariamidae, both distributed in South America. They are found on fairly dry open country, grasslands or scrubs. Ecologically they are the South American counterpart of the raptorial bird of Africa, the Secretary Bird.

Seriemas are charismatic and are thought to be the closest living relatives of the Phorusrhacids (which are known from fossils from Africa, South and North America). The Paleogene Idiornithids and Bathornithids from Europe and North America respectively are possibly related too. All these birds are Cariamiformes according with Degrange (2012) and in this sense, the records of Itaboraí are within the oldest representative of this order.

Based on a coracoid and two humeri, Mayr et al. (2011b) nominated the taxon *Itaboravis elaphrocnemoides* which presents morphological similarities with the European taxon *Elaphrocnemus*, although the humerus also share some features with the South American tinamous (Tinamidae), such as the weakly developed crista deltopectoralis. This last feature would indicate a limited ability to fly. There is some additional material (carpometacarpus and tibiotarsus) possibly belonging to the same species, so it seems valid to assume that *Itaboravis* was very much abundant in Itaboraí (Mayr et al. 2011a).

Paleopsilopterus itaboraiensis was nominated on the base of a very fragmentary right tarsometatarsus and two tibiotarsi severely deformed. Alvarenga (1985a) supports the inclusion within Phorusrhacidae Psilopterinae based on the pons supratendineus of the tibiotarsus is transversely oriented. We disagree because the pons is oblique in all known phorusrhacids. Also, the eminentia intercotylaris wide, rounded, and poorly extended proximally of *Paleopsilopterus* is conspicuously different of all known Psilopterinae, a condition that had already been noted by Alvarenga and Höfling (2003). Agnolín (2009) supports the exclusion of *Paleopsilopterus* from the Phorusrhacidae and its inclusion, with no quite convincing arguments, in the European family Idiornithidae, which are also considered as extinct relatives to the Seriemas. Due to the fragmentary state of the material and another set of characters that *Paleopsilopterus* shared with other Cariamiformes, here we prefer to consider it as belonging to an uncertain family of Cariamiformes (Cariamiformes *incerti familiae*).

In the 1990s, *Eutreptodactylus itaboraiensis* was published based on a fragmentary tarsometatarsus and considered as the oldest and most primitive Cuculidae (Baird and Vickers-Rich 1997). Features of the trochlea metatarsi II and IV of the tarsometatarsus seemed to indicate that it was a small zygodactyl cuckoo. Nevertheless, Mayr et al. (2011) believe that this hypothesis is weak. Nowadays the holotype is lost and obviously, the revision in hand is not possible. Additional material (a distal left tibiotarsus) was tentatively assigned to this taxon by Mayr et al. (2011).

Indeterminated birds from Itaboraí are represented by a carpometacarpous and fragmentary tibiotarsi (Mayr et al. 2011a).

5.2 Eocene Birds

Records of continental birds from the Eocene of South America are actually very rare. Some few remains mentioned below are from localities in Chile and Argentina.

Presbyornithidae was an extinct family of waterbirds with an apparently global distribution that lived until the Oligocene (see Kurochkin et al. 2002 who summarized the worldwide fossil record). Its place within Anseriformes is not now under discussion. The frequency of these birds in aquatic and semi-aquatic environments by the earliest Paleogene is absolutely sustained. Because presbyornithids are also well documented from the late Cretaceous, they provide a clue to selective avian survivorship across the Cretaceous-Paleogene boundary (Kurochkin and Dyke 2010).

Howard (1955) was the first to recognize the presence of presbyornithids in South America. Based on a skeleton unearthed from the Sarmiento Formation in the Eocene Cañadón Hondo locality (near to Paso Niemann, Chubut, Argentina), she nominates the species *Telmabates antiquus*. Later, Cracraft (1970) describes a new species, *T. howardae* based on a distal end of a tibiotarsus from the same locality and age. A later review (Ericson 2000) excludes this taxon of the family, mainly by the morphology of the condylus lateralis (rounded in *T. howardae* but ovoid in all known Presbyornithids). In his classic synthesis of the avian fossil record of Argentina, Tonni (1980) refers to these same findings but synonymized *T. howardae* with *Presbyornis pervetus*. *Presbyornis* is a well-known fossil presbyornithid that was approximately the size and shape of a goose, but with longer legs; judging from numerous fossil findings, supposedly nested in colonies around shallow lakes.

A carpometacarpus of a presbyornithid from Vaca Mahuida Formation (late Paleocene-middle Eocene) was also described by Tambussi and Noriega (1998). The formation is exposed at Sierra El Fresco, southeast of Puelen locality

(La Pampa, Argentina). A shallow, brackish-water environment is inferred from the deposits. Features like a straight carpometacarpus, with major and minor metacarpals parallel, presence of large scars for the insertions of lig. ulnocarpometacarpale dorsale, and radiocarpometacarpale dorsale, are diagnostic for Presbyornithidae (Ericson 2000).

From Cañadón Vaca, Cracraft (1971) described *Onychopteryx simpsoni* placed within the monotypic taxon Onychopterygidae. The material from which the taxon was nominated is a proximal end of a right tarsometatarsus, badly preserved. It was discovered in the 1930s by Simpson during the Scarrit paleontological expeditions conducted by the American Museum of Natural History in Patagonia. Brodkorb (1978) considered the material too fragmentary to support any affinity with other birds, an approach taken by further authors.

Phorusrhacids Psilopterinae has been reported from the same place, Cañadón Vaca (Tonni and Tambussi 1986) which is the oldest record of the family. The family was also been reported from the late Eocene Gran Hondonada locality (Chubut, Argentina) by Acosta Hospitaleche and Tambussi (2005) (Fig. 5.2). The systematic placement of the specimen is now in conflict. While Agnolín (2009) indicates that it would be a new species of Idiornithidae, Degrange (2012) pointed out that the poor development of the eminentia intercotylaris it is indicative that this taxon it is not a Phorusrhacidae. Without any new material the systematic position cannot be certified.

In a revision of the fossil record of Accipitridae, Agnolín (2006a) assign to this family some fragmentary ungual phalanxes (MPEF 1050 y MLP 74-II-1-21) proceeding from the Eocene of Gran Hondonada. However, the presence of lateral grooves that characterize these remains dismisses this assignation.

To our knowledge, the only other Eocene non-penguin records of bird were recovered recently from the Algarrobo unit (middle to late Eocene), central Chile.

(a)   **(b)**

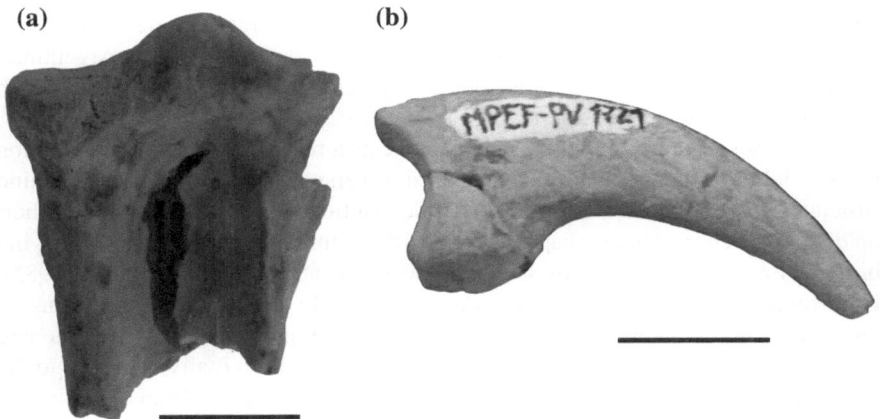

Fig. 5.2 Cariamiformes from the Eocene Gran Hondonada locality: **a** Tarsometatarsus MPEF-PV1722, **b** Ungual phalanx MPEF-PV1721

Remains of a tibiotarsus assigned to Ardeidae (Sallaberry et al. 2010) and unassociated proximal and distal fragments of right femur assigned with doubts to Procellariiformes (Yury-Yáñez et al. in press) invite to continue the search at the same deposits.

5.3 Oligocene Avian Taxa Proposed by Ameghino: New and Old Systematic Hypotheses

Toward the end of the nineteenth century, Florentino Ameghino (1854–1911) had become a distinguished Argentine paleontologist. His publications applied concepts and conclusions with a paleontological approach applied to evolutionary biology on the fossil evidence that placed him among the few world figures of that time. Many of these concepts and theories are overturned in his theoretical work "Filogenia" published in 1884. Although his paleontological activity was primarily devoted to mammals, his contributions were also very important in geology, stratigraphy, archaeology, and philosophy.

In the field of birds, his activity was not lower. In only ten works published between 1882 and 1905, he nominated 81 species from a few Oligocene and Miocene localities of Patagonia and Pleistocene of the Pampas (Table 5.1). Most of them were collected by his brother Carilos Ameghino (1865–1936) who between 1887 and 1902, made more than a dozen trips to Patagonia (some more than 1 year). Interestingly, except for one species (*Andrewsornis abbotti*), there are not other Oligocene birds from Argentina than those described by Ameghino more than 100 years ago.

In 1895, Ameghino compiled in one "Boletín del Instituto Geográfico Militar Argentino" the results of his work about fossil birds of Patagonia. In this publication he nominated the most striking species of all he recognized: *Phorusrhacos longissimus* Ameghino 1887 from the Miocene of Santa Cruz province. This work, together with the "Enumeración de las aves fósiles de la República Argentina" (Ameghino 1891a), remains as an obligatory reference for those engaged with paleornithology of South America.

In all cases, the specimens are macroscopic as a result of the style of collection in the field of the epoch, reflecting bird diversity only partially. Ameghino structured his arguments with meticulous descriptions but generally stopped short on diagnostic characters, perhaps because of their limited experience in identifying birds and a limited number of skeletons available for comparison (Olson 1985). Unfortunately, his descriptions are not accompanied by appropriate illustrations. In this sense, contrasts sharply with the catalog of fossil birds written by Moreno and Mercerat (1891) which well illustrates specimens which originally belonged to the collection that Ameghino yielded at the Museo de La Plata in 1886.

Also, many of the Ameghino's species were based on extremely fragmented and poorly preserved material (e.g., Ameghino 1895, 1899). Indeed in the context of

Table 5.1 List of nominated birds by Ameghino between the years 1882 and 1905 with the taxonomic status and the corresponding reference

	Species	Status	Systematic	Years	References
1	*Aminornis excavatus*	Valid	Anseriformes	1899	I
2	*Anisolornis excavatus*	Valid	Gruiformes	1891	U
3	*Anissolornis excavatus*	2		1899	B
4	*Apterodytes ictus*	32		1901	R
5	*Argyrodyptes microtarsus*	Valid	Procellariiformes Procellariidae?	1905	C
6	*Arthrodytes andrewsi*	Valid	Spheniscidae	(1901)	H, K
7	*Arthrodytes andrewsi*	6		1905	H, K
8	*Arthrodytes grandis*	48		1905	H, K, L
9	*Asthenopterus minutus*	*Lagopterus minutus*	Falconidae Polyborinae	1891	C, D
10	*Aucornis euryrhynchus*	63		1899	G
11	*Aucornis solidus*	Nomen dubium	incertae sedis	1899	J, U
12	*Badiostes patagonicus*	Valid	Falconidae	1895	C
13	*Brontornis platyonyx*	*Brontornis burmeisteri*		1895	G
14	*Callornis giganteus*	62		1895	G
15	*Chenalopex debilis*	*Neochen debilis*		1891	B
16	*Ciconiopsis antarctica*	Nomen inquirendum	Phorusrhacidae	1899	U
17	*Cladornis pachypus*	Valid	Aves Neornithes incerti ordinis	1895	U, P
18	*Climacarthrus incompletus*	Nomen dubium		1899	U, P
19	*Cruschedula revola*	Valid	Aves incerti ordinis	1899	U, P
20	*Eoneornis australis*	Valid	Anseriformes incertae familiae	1895	F, N
21	*Eucallornis giganteus*	62		1901	G
22	*Euelornis patagonicus*	Valid	Anseriformes incertae familiae	1895	N
23	*Isotremornis nordenskjöldi*	48		1905	H, K, L
24	*Liornis floweri*	62		1895	G
25	*Liptornis hesternus*	Valid	Pelecaniformes Anhingidae	1895	T
26	*Loncornis erectus*	Nomen dubium	Mammalia	1899	M
27	*Lophiornis obliquus*	Valid	Anseriformes	1891	F
28	*Loxornis clivus*	Valid	Anseriformes	1895	F
29	*Metancylornis curtus*	48		1905	H, K, L

(continued)

Table 5.1 (continued)

	Species	Status	Systematic	Years	References
30	Neculus rothi	Nomen dubium		1905	U, O
31	Opistodactylus patagonicus	Valid	Rheidae	1891	C, E
32	Palaeoapterodytes ictus	Nomen dubium		1905	R
33	Palaeospheniscus affinis	Palaeospheniscus bergi		1895	H, K, L
34	Palaeospheniscus concavus	Nomen dubium		1905	H
35	Palaeospheniscus convexus	Nomen dubium		1905	H
36	Palaeospheniscus gracilis	Palaeospheniscus bergi		1899	H, K, L
37	Palaeospheniscus intermedius	Palaeospheniscus patagonicus		1905	H, K, L
38	Palaeospheniscus interruptus	Palaeospheniscus bergi		1905	H, K, L
39	Palaeospheniscus medianus	Palaeospheniscus bergi		1905	H, K, L
40	Palaeospheniscus nereius	Palaeospheniscus bergi		1901	H, K, L
41	Palaeospheniscus planus	Palaeospheniscus bergi		1905	H, K, L
42	Palaeospheniscus robustus	Palaeospheniscus patagonicus		1895	H, K, L
43	Palaeospheniscus rothi	Palaeospheniscus bergi		1905	H, K, L
44	Palaeospheniscus wimani	Nomen nullum		1905	H
45	Paraptenodytes andrewsi	6		1901	H, J, K
46	Paraptenodytes curtus	48		1901	H, K, L
47	Paraptenodytes grandis	48		1901	H, K, L
48	Paraptenodytes robustus	Valid	Spheniscidae	(1895)	H, K, L
49	Paraspheniscus bergi	Palaeospheniscus bergi		1905	H, K, L
50	Paraspheniscus nereius	Palaeospheniscus bergi		1901	H, K, L
51	Pelecyornis minutus	Psilopterus bachmanni		1891	G
52	Pelecyornis tubulatus	Psilopterus lemoinei		1895	G
53	Perispheniscus robustus	48		1895	H, K
54	Perispheniscus wimani	Palaeospheniscus biloculata		1905	H, K, L
55	Phororhacos affinis	Psilopterus affinis		1899	G
56	Phororhacos delicatus	Psilopterus bachmanni		1891	G
57	Phororhacos inflatus	Patagornis marshi		1891	G
58	Phororhacos longissimus	62		1889	G

(continued)

Table 5.1 (continued)

	Species	Status	Systematic	Years	References
59	*Phororhacos modicus*	*Psilopterus lemoinei*		1895	G
60	*Phororhacos platygnathus*	62		1891	G
61	*Phororhacos shenensis*	62		1891	U
62	*Phororhacos sehuensis*	63		1891	G
63	*Phorusrhacos longissimus*	Valid	Phorusrhacidae	1887	G
64	*Physornis fortis*	Valid	Phorusrhacidae	1895	G
65	*Prociconia lydekkeri*	*Ciconia lydekkeri*	Ciconiiformes Ciconiidae	1891	A, S
66	*Protibis cnemialis*	Valid	Ciconiiformes Threskiornithidae?	1891	C, D
67	*Pseudolarus eocaenus*	*Psilopterus lemoinei*		1891	G
68	*Pseudolarus guaraniticus*	Nomen dubium		1899	U, P
69	*Pseudospheniscus concavus*	Nomen dubium		1905	H
70	*Pseudospheniscus convexus*	Nomen dubium		1905	H
71	*Pseudospheniscus interplanus*	Nomen nullum		1905	H
72	*Pseudospheniscus planus*	*Palaeospheniscus bergi*		1905	H, K
73	*Rhea fossilis*	Valid	Rheidae	1882	C, E
74	*Riacama caliginea*	Valid	Aves Neornithes incerti ordinis	1899	P, G
75	*Smiliornis penetrans*	Valid	Aves Neornithes incerti ordinis	1899	U
76	*Teleornis impressus*	Valid	Anseriformes Anatidae Tadornini	1899	F
77	*Thegornis debilis*	Valid	Falconidae	1895	Q
78	*Thegornis musculosus*	Valid	Falconidae	1895	Q
79	*Tiliornis senex*	Valid	Phoenicopteridae?	1899	F
80	*Tolmodus inflatus*	*Patagornis marshi*		1891	G
81	*Treleudytes crassa*	*Palaeospheniscus patagonicus*		1905	H, K, L

In the second column, the species are arranged alphabetically. The numbers in the Status column refer to the taxon which is junior synonym and correspond to the numbers in the first column. The names of the Status column indicates which species is junior synonym. In the Ref. column, main bibliographic references are included. (A) Brodkorb (1963), (B) Brodkorb (1964), (C) Tonni (1980), (D) Tonni and Tambussi (1986), (E) Tambussi (1995), (F) Tambussi and Noriega (1996), (G) Alvarenga and Höfling (2003), (H) Acosta Hospitaleche (2004), (I) Agnolín (2004), (J) Agnolín (2006a), (K) Acosta Hospitaleche et al. (2007), (L) Acosta Hospitaleche and Tambussi (2008), (M) Agnolín (2008), (N) Worthy (2008), (O) Acosta Hospitaleche (2009), (P) Mayr (2009), (Q) Noriega et al. (2011), (R) Acosta Hospitaleche (2010), (S) Agnolín (2010), (T) Cenizo and Agnolín (2010), (U) This work. The numbers in the Year column correspond to the year in which Ameghino nominated the species. The Systematic column indicate the closest location possible to the date, according to the reference. Suprageneric terminology follows Matthews (1973) the contemporary paleornithology, many of these materials have not been assigned beyond the ordinal level or even the class is doubtful. The ICZN is very clear on the actions to follow when a taxa nominated is under review, and about the caution in the process of nomination of new taxa (de la Fuente 2005).

Over the past 30 years, many Ameghino's species have been reassessed. In many cases, the revisions have been made on the study of the original material but more frequently, only on the original descriptions or figures. As a result of this, many of synonymies or new combinations established have been wrong (Table 5.1).

As a consequence of rearrangement of collections made during the last decade both in the British Museum (BMNH) and Museo Argentino de Ciencias Naturales (MACN), some holotypes that were considered lost, have been located. Only two of the original materials on which Ameghino founded his species have not been located at the collections (*Tiliornis senex* and *Loxornis clivus*). We cannot deny that many of the birds studied by Ameghino are crucial in the reconstruction of the evolutionary history of the biota of South America. It is a prerequisite for any study involving paleoenvironmental and paleobiogeographic inferences to have a clear idea of the systematic diversity. Still, it is clear to us that a profound repositioning of the Ameghino's species could only be valid in the light of the discovery of new materials. Most of the names given by Ameghino are for penguins (33 of the 81 given names). They were studied in depth by Hospitaleche (2004, 2007, 2009, 2010; Acosta Hospitaleche and Tambussi 2008) and are not the purpose of this work.

Next, we will explore the Oligocene continental birds that Ameghino named. Later we will do the same with the Miocene.

We mentioned earlier that most of the Oligocene birds that are known from South America came from Argentina (the others are from Brazil), and most of them were described by Ameghino (1895, 1899). It is worthy to mention here that revising American collections, one of the authors (FJD) found new unpublished material proceeding from the Cabeza Blanca locality. These materials are under study by FJD and CPT.

Remains are fragmentary and of dubious or debated assignment (Fig. 5.3). Agnolín (2004) made an attempt to review these materials, but his arguments of changes are unconvincing, some of the descriptions are confusing and the figures provided are deficient.

For example, *Riacama caliginea* from the "Deseado Formation" ("Formación Guaranítica" sensu Ameghino) at Santa Cruz (Argentina) was considered as a Phorusrhacidae Psilopterinae by Brodkorb (1967), a Cariamidae by Tonni (1980) and Agnolín (2004), and an Idiornithidae by Agnolín (2009). However, the species is based on a fragment of shaft and sternal extremity of a right coracoid (Fig. 5.3a) that provides limited information and does not allow to infer its exact affinities (Alvarenga and Höfling 2003).

As *Riacama caliginea*, *Aminornis excavatus* was exhumed from the "Deseado Formation" at Santa Cruz. It was diagnosed from a proximal fragment of a right coracoid (Fig. 5.3b). Originally (Tonni 1980) assigned the species to the aramids

(Gruiformes) but later it was relocated within the Anseriformes (Agnolín 2004), a criterion with which we agree. It is considered prudent to retain *Aminornis excavatus* as a valid species of Anseriformes.

Opinions toward the distal end of right tibiotarsus of *Loxornis clivus* are dissimilar. Some early authors such as Loomis (1914) suggested that it would have affinities with *Psilopterus* while others (Alvarenga 1999; Cenizo and Agnolín 2010) with Anhimidae (the family of Anseriformes that includes the screamers). Because the holotype could not be relocated in the MACN nor BMNH collections, we studied the calcotype deposited at the Field Museum of Natural History in Chicago (FM PA 20, Fig. 5.3c). General morphology of the tibiotarsus, central position of the supratendinous bridge and the presence of a medial ridge at the canal extensorius allow to consider *Loxornis clivus* as a valid species of Anseriformes as it was suggested by Ameghino, but insufficient to make a more accurate systematic position. Additional materials (humerus, coracoid, sternum, femur, tibiotarsus, fibula, tarsometatarsus, and some phalanges) from the west of Puerto Visser (Chubut, Argentina) were assigned to this species (Loomis 1914). This assignment was subsequently rejected (Patterson 1941) and the materials were assigned to other birds such us *Smiliornis* and *Andrewsornis* or even to ungulate mammals.

Smiliornis penetrans from the "Deseado Formation" of Santa Cruz was found on a very small proximal portion of left coracoid (Fig. 5.3d). Patterson (1941) and Tonni (1980) assigned it to a Phorusrhacidae Psilopterinae while Alvarenga and Höfling (2003) based on the poor diagnostic nature of the material, suggest that it may be synonymous of *Psilopterus affinis*. *Smiliornis* is grouped with the Idiornithidae by Agnolín (2009). However, characters preserved in the fossil do not provide clarity on its affinities.

In 1899, Ameghino founded on the basis of a highly eroded tarsometatarsus fragment (Fig. 5.3e) the species of Falconidae *Climacarthrus incompletus* also from the "Deseado Formation" of Santa Cruz. Later, the species was assigned to Accipitriformes Accipitridae by Brodkorb (1964), criterion followed by Agnolín (2004). Mayr (2009) suggests that the final place depends on the discovery of more complete specimens. We believe that the morphology of the preserved fragment do not allow any kind of assignation. For this reason, we propose that *Climacarthrus incompletus* must be considered as *nomen dubium*.

Teleornis impressus diagnosed on a distal portion of right humerus (Fig. 5.3f) has undoubtedly features of Anseriformes Anatidae (Mayr 2009). According to Agnolín (2004) and Cenizo and Agnolín (2010) it would be an Anatidae Tadornini. As *Riacama caliginea* the specimen was exhumed from the "Deseado Formation" at Santa Cruz.

Cruschedula revola was originally located in its own and new family, Cruschedulidae. Later authors reassigned it to the Accipitridae (Tonni 1980; Agnolín 2006a). It is based on a small fragment of a proximal right scapula (Fig. 5.3g) that does not allow to inferred reliable phylogenetic affinities (Mayr 2009). *Cruschedula revola* should be considered as a valid species of an uncertain Order (Aves *incerti ordinis*).

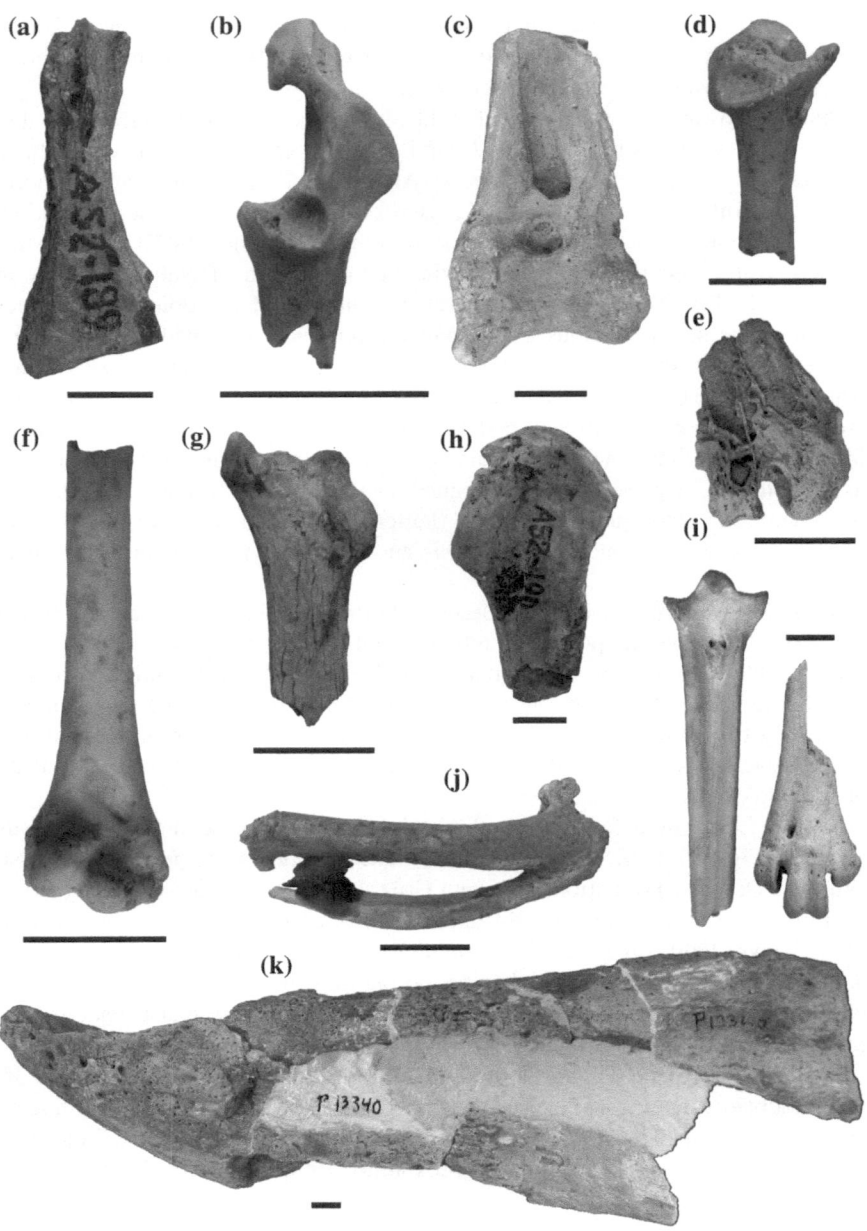

◄ **Fig. 5.3** Oligocene fossil birds: **a** *Riacama caliginea*, proximal portion of right coracoid (MACN A52-189), **b** *Aminornis excavatus*, distal portion of right coracoid (MACN A10305), **c** *Loxornis clivus*, distal portion of tibiotarsus (FM-PA 20), **d** *Smiliornis penetrans*, distal portion of left coracoid (MACN A52-183), **e** *Climacarthrus incompletus*, distal portion of fragmentary tarometatarsus (MACN A11667), **f** *Teleornis impressus*, distal portion o right humerus (MACN A108-89), **g** *Cruschedula revola*, proximal fragment of right scapula (MACN A11039), **h** *Pseudolarus guaraniticus*, proximal fragment of humerus (MACN A52-190), **i** *Psilopterus affinis*, right tarsometatarsus (MACN A52-184), **j** *Ciconiopsis antarctica*, fragmentary left carpometacarpous (MACN A11666), **k** *Physornis fortis*, fragmentary mandibule (FM-P13340). Scale bar = 1 cm

Also in the "Deseado Formation" of Santa Cruz, Ameghino recovered an extremely fragmentary humerus (Fig. 5.3h) that his brother named as *Pseudolarus guaraniticus* and assigned to Phorusrhacidae. Recently, Agnolín (2006b) relocates this taxon within Psilopterinae and proposed the new combination *Psilopterus guaraniticus*. However, the specimen is considerably larger than any known species of *Psilopterus*, and the material does not permit a reliable assignment because of its preservation. Also, following the recommendations of the ICZN, we propose here that *Pseudolarus guaraniticus* to be considered as a *nomen dubium*.

Ameghino founded the species *Ciconiopsis antarctica* based on a fragment of left carpometacarpus (not right as shown in Agnolín 2004, Fig. 5.3) from the "Deseado Formation" which assigns to Ciconiidae (Ciconiiformes, storks and allies) (Fig. 5.3j). The species was removed from the Ciconiiformes and located within Phorusrhacidae Psilopterinae (Agnolín 2004) and, moreover, was considered synonym of *Psilopterus* (Mayr 2009). It is difficult to understand why Agnolín (2004) described the processus alularis 90° disposed as diagnostic of Psilopterins ("...proceso alular a 90 grados con respecto al eje de la diáfisis del metacarpal mayor", Agnolín 2004), p 243, when until recently is not possible to describe this feature in any Psilopterinae. Few carpometacarpi of undoubtedly Psilopterinae are at the museum collections (YPM-PU 15402, BMNH A559-1, MPM-PV4243) and only in one case the processus alularis is preserved. Degrange et al. (2011) described the carpometacarpous of MPM-PV4243, a new material assigned to *Psiloterus bachmanni* and in this taxon the processus alularis is oblique disposed (not perpendicular). In the other materials, this feature is not preserved. Also, the torsion and relative development of the os metacarpale minus and the dorsoventral extension of spatium intermacarpalis in *Ciconiopsis* looks more like a Patagornithinae feature than a Psilopterinae one. In short, characters preserved allow the assignment to Phorusrhacidae. Only two taxa from the early Oligocene of Phorusrhacidae are known, *Andrewsornis abotti* Patterson 1941 (fragment of skull and mandible, femur and phalanges) and *Psilopterus affinis* (Ameghino 1899) (tarsometatarsus with the middle portion of the shaft lacking, Fig. 5.3i). These species are based on bones not homologous with those assigned to *Ciconiopsis antarctica*. The category proposed by the ICZN for cases like this is *nomen inquirendum* (Mones 1989).

In his work of 1899, Ameghino also nominated *Phororhacos affinis* based on a fragmentary right tarsometatarsus recorded in late Oligocene deposits at Golfo San Jorge basin, (southern Patagonia, Argentina). Patterson (1941) considered it as synonymous of *Smiliornis penetrans*, but in a later review, Alvarenga and Höfling (2003) propose that this species belongs to *Psilopterus'* genus and proposed the new combination *Psilopterus affinis* and therefore, this is the smallest species of a phorusrhacid Psilopterinae.

Physornis fortis from Santa Cruz (Argentina) is another Oligocene species nominated by Ameghino. It was a giant bird included in the Phorusrhacids Physornithinae in the sense of Agnolín (2006b, 2009). Besides its size, features such as high and short mandibular symphysis (Fig. 5.3k), and robust, broad and relatively short tarsometatarsus, characterized this taxon.

Agnolín (2008) is not wrong when states that *Loncornis erectus,* based on a fragment of humerus of a juvenile (and not a femur as indicated by Mayr 2009) is a mammal.

The coracoid referred to *Tiliornis senex* could not be located in MACN and MLP collections; its original systematic position within Phoenicopteridae cannot be corroborated.

Also from the "Deseado Formation" of Santa Cruz and Late Oligocene in age, was described *Cladornis pachypus*. Of all the birds recognized by Ameghino, the affinities of this species are the most controversial. The holotype, a right tarso-metatarsus (BMNH A589), was figured by the very first time by Mayr (2009). It is short, dorsoventrally flat and robust, with a large facet for a digit I placed very proximal, and trochleas for digits II and IV located in the same horizontal plane. Tonni (1980) placed *Cladornis* within the Pelecaniformes and Olson (1985) inferred that it was zygodactylous. We agree with Mayr (2009) in that although the material is very distinctive, its adequate phylogenetic affinity requires the discovery of more complete materials.

Additionally to the taxa nominated by Ameghino, the Oligocene *Andrewsornis abbotti* was established by Patterson in 1941 to describe the largest and single Patagornithinae representative of the entire Paleogene. The taxon was described based on an incomplete skull, jaw, incomplete coracoid, and phalanges of digit II. Later, they were referred to the same species additional materials (femur and jaw). Agnolín (2009) states that Patagornithinae does not constitute a natural group and that *Andrewsornis* would correspond to a basal form within Phororhacoidea Phorusrhacidae unrelated with *Andalgalornis* and *Patagornis*. Today, the coracoid and phalanges could not be located in their original repositories.

5.4 Oligocene Avian Taxa from the Brazilian Tremembé Formation

The phorusrhacid *Paraphysornis,* the vulture *Brasilogyps,* the teratornithid *Taubatornis,* the hoatzin *Hoazinavis,* the anseriform *Chaunoides,* two species of the galliform *Ameripodius,* and two species of flamingoes constitutes the bird association of the Tremembé Formation (Late Oligocene to Early Miocene) of at Taubaté Basin, Brazil. It was earlier mentioned that the deposits are of lacustrine origin, and it could have been a marshy environment goberned by periods of water shortage at the time of deposition (Olson and Alvarenga 2002; Alvarenga and Höfling 2003).

Paraphysornis brasiliensis is the best Physornithinae (Phorusrhacidae) represented and the only one present in Tremembé. The partially complete skeleton only lacks the skull, pelvis, and sternum. The short, robust tarsometatarsus with wide trochlear spread of *Paraphysornis brasiliensis* together with its body mass estimated on 180 kg (Alvarenga and Höfling 2003) undoubtedly indicates a terrestrial habit. The general morphology of the tarsometatarsus does not differ much with that of the mihirung *Dromornis stirtoni* (Dromeornithiformes). Murray and Vickers-Rich (2004) argued that high masses do not impose limits on the cursorial ability of the Australians mihirung and that even those for which estimated masses are of 500 kg, were able to run. This idea contrasts with Alvarenga (1982) who argues that *Paraphysornis* would only have slow-moving capacity. An important element for establishing locomotor and postural habits is the pelvis (Degrange 2012), which unfortunately has not been recovered.

Cathartidae is an ancient group of birds in the South American Cenozoic. The oldest record corresponds to *Brasilogyps faustoi* from the Oligocene of Brazil (Tambussi and Noriega 1996; Alvarenga et al. 2008). *Brasilogyps* had a bigger size than the new world vulture *Coragyps* (Feduccia 1999) and was described on the base of a distal end of a right tibiotarsus and a proximal fragment of a right tarsometatarsus (Alvarenga 1985b).

A distal end of a right tibiotarsus and a fragment of ulna were referred to a new genus and species, *Taubatornis campbelli* which is the smallest and oldest teratorn (Teratornithidae? Ciconiiformes) til date (Olson and Alvarenga 2002). Teratorns were giant volant birds, supposedly carnivorous or scavengers, identified from the Pleistocene of the Americas (North, South and Cuba) and Miocene of South America. Teratorns and vultures share in Tremembé the scavenger niche as has also been reported for other sites.

Hoazinavis lacustris is the oldest and smaller Opisthocomiformes (Mayr et al. 2011b) from which are known the humerus, scapula and coracoides, independent of each other unlike the living hoazin *Opisthocomus hoatzin.* Living hoatzins are found in forests of northern South America, especially along rivers and streams. Fossils of Opisthocomiformes were recovered in the early late Miocene of Namibia (Africa) which documents that the extant Neotropic distribution of hoatzins is relictual (Mayr et al. 2011b) and provides one example of transatlantic rafting among birds. A more modern member of this order, *Hoazinoides magdalenae* is known from fragmentary remains, including the back portion of the skull and other leg bones coming from the middle Miocene of Villavieja Formation from Colombia.

Chaunoides antiquus is the only published Paleogene remain of Anhimidae. Based on several isolated postcranial elements, it was described by Alvarenga (1999). This extinct species is still smaller than the smallest of the living Anhimidae, the northern Screamer *Chauna chavaria*.

Only two Galliform records were described from the Tremembé Formation (Alvarenga 1988, 1995), both currently recognized as Quercymegapodiidae and belonging to the genus *Ameripodius: Ameripodius granivora* (= *Taubacrex granivora* following Mourer-Chauviré 2000, originally described as a Rallidae by Alvarenga 1988) and *Ameripodius silvasantosi*. Quercymegapodiidae is a primitive clade of Galliformes that resembles the stocky medium-large chicken-like megapodes or mound-builders. They have also been identified in the middle Eocene and Lower Miocene of France (Mourer-Chauviré 2000; Lindow and Dyke 2007). *Ameripodius* emphasize the similarity between the European and South American avifaunas for the earliest Cenozoic times (Mourer-Chauviré 2000).

Flamingos (Phoenicopteriformes) of Tremembé are members of two different families: Phoenicopteridae and Palaelodidae (Alvarenga 1990). They are represented by *Agnopterus sicki* and *Palaelodus aff. ambiguous* respectively. The first one, from which is only known a distal end of a tibiotarsus, is grouped together with living flamingos, the well known wading birds of cosmopolitan distribution that live associated with marshes. The extinct Paleolodids from the early Tertiary of Europe, both Americas, and Australia (Olson and Feduccia 1980; Cheneval 1983; Alvarenga 1990; Boles 1991) have been described as swimmers (Olson and Feduccia 1980) or even divers (Cheneval and Escuillié 1992 versus Mayr 2004) and filter-feeder aquatic birds.

References

Acosta Hospitaleche C (2004) Los pingüinos (Aves, Sphenisciformes) fósiles de Patagonia. Sistemática, biogeografía y evolución. Dissertation, Universidad Nacional de La Plata, p 321

Acosta Hospitaleche C (2007) Revisión sistemática del género y especie Palaeospheniscus biloculata nov. comb. (Aves, Spheniscidae) de la Formación Gaiman. Ameghiniana 44:417–426

Acosta Hospitaleche C (2009) Estatus taxonómico de *Neculus rothi* (Aves; Sphenisciformes) del Mioceno temprano de Patagonia, Argentina. Ameghiniana 46:199–201

Acosta Hospitaleche C (2010) Taxonomic status of *Apterodytes ictus* Ameghino, 1901 (Aves; Sphenisciformes) from the Early Miocene of Patagonia, Argentina. Neu Jah Geol Pal-Ab 255:371–375

Acosta Hospitaleche C, Tambussi C (2005) Phorusrhacidae Psilopterinae (Aves) en la Formación Sarmiento de la localidad de Gran Hondanada (Eoceno Superior), Patagonia, Argentina. Rev Esp Pal 20:127–132

Acosta Hospitaleche C, Tambussi CP (2008) South American fossil penguins: a systematic update. Oryctos 7:109–127

Agnolín FL (2004) La posición sistemática de algunas aves fósiles deseadenses (Oligoceno Medio) descriptas por Ameghino en 1899. Rev Mus Arg Cienc Nat 6:239–244

Agnolín FL (2006a) Notas sobre el registro de Accipitridae (Aves, Accipitriformes) fósiles argentinos. Stud Geol Salmant 42:67–80

Agnolín FL (2006b) Posición sistemática de algunas aves fororracoideas (Ralliformes; Cariamae) Argentinas. Rev Mus Arg Cienc Nat 8:27–33

Agnolín FL (2008) Reconsideración sobre la posición sistemática de *Loncornis erectus* Ameghino, 1899 (Mammalia; non Aves). Stud Geol Salmant 44:9–12

Agnolín FL (2009) Sistemática y filogenia de las aves fororracoideas (Gruiformes: Cariamae). Monografías Fundación Azara, p 79

Agnolín F (2010) El registro fósil de "Ciconia lydekkeri" Ameghino, 1891 en el Pleistoceno de sud América. Studia Geologica Salmanticensia 45:53–58

Alvarenga HMF (1982a) Uma gigantesca ave fóssil do Cenozóico brasileiro: *Physornis brasiliensis* sp. n. An Acad bras Ciênc 54:697–712

Alvarenga HMF (1982b) Aves: Phorusrhacidae. An Acad Bras Ciênc 65:403–406

Alvarenga HMF (1983) Uma ave ratite do Paleoceno brasileiro: bacia calcária de Itaboraí, estado do Rio de Janeiro, Brasil. Bol Mus Nac do Rio de Jan, Geol 41:1–47

Alvarenga HMF (1985a) Um novo Psilopteridae (Aves: Gruiformes) dos sedimentos Terciários de Itaboraí, Rio de Janeiro, Brasil. Anais do Cong Bras de Pal 8. Série Geol 27:17–20

Alvarenga HMF (1985b) Notes on the Cathartidae (Aves) and description of a new genus from the Brazilian Cenozoic. An Acad Bras Ciênc 57:349–357

Alvarenga HMF (1988) Ave fóssil (Gruiformes: Rallidae) dos folhelhos da Bacia de Taubaté, Estado de São Paulo, Brasil. An Acad Bras Ciênc 60:321–332

Alvarenga HMF (1990) Flamingos fosseis da bacia de Taubaté, estado de São Paulo, Brasil: desçricao de nova especie. An Acad bras Ciênc 62:335–345

Alvarenga HMF (1993) *Paraphysornis* novo gênero para *Physornis brasiliensis*

Alvarenga HMF (1995) A large and probably flightless anhinga from the Miocene of Chile. Cou For Senck 181:149–161

Alvarenga HMF (1999) A fossil screamer (Anseriformes: Anhimidae) from the middle tertiary of Southeastern Brazil. Smith Contrib Paleobiol 89:223–230

Alvarenga HMF (2010) *Diogenornis fragilis* Alvarenga, 1985, restudied: a South American ratite closely related to Casuariidae. 25th Int Ornit Cong, p 143

Alvarenga HMF, Höfling E (2003) Systematic revision of the Phorusrhacidae (Aves: Ralliformes). Pap Av Zool 43:55–91

Alvarenga HMF, Guilherme RR, Brito RM, Hubbe A, Höfling E (2008) *Pleistovultur nevesi* gen. et sp. nov. (Aves: Vulturidae) and the diversity of condors and vultures in the South American Pleistocene. Ameghiniana 45:613–618

Ameghino F (1891) Enumeración de las aves fósiles de la República Argentina. Rev Arg Hist Nat 1:441–453

Ameghino F (1895) Sobre las aves fósiles de Patagonia. Bol Inst Geog Arg 15:501–602

Ameghino F (1899) Sinopsis geológico-paleontológica de la Argentina. Segundo Censo Rep Arg 1:112–255

Baird RF, Vickers-Rich P (1997) *Eutreptodactylus itaboraiensis* gen. et. sp. nov., an early cuckoo (Aves: Cuculiformes) from the Late Paleocene of Brazil. Alcheringa 21:123–127

Boles WE (1991) The origin and radiation of Australasian birds: perspectives from the fossil record. In: Cossee RO, Flux JEC, Heather BD, Hitchmough RA, Robertson CJR, Williams MJ (eds) Bell BD. Acta XX Cong Int Ornit, Christchurch

Brodkorb P (1963) A giant flightless bird from the Pleistocene of Florida. Auk 80:111–115

Brodkorb P (1964) Catalogue of fossil birds. Part II (Anseriformes through Galliformes). Bull Fla State Mus Biol Sci 8:195–335

Brodkorb P (1967) Catalogue of fossil birds, Part III (Ralliformes, Ichthyornithiformes, Charadriiformes). Bull Fla State Mus Biol Sci 2:99–220

Brodkorb P (1978) Catalogue of fossil birds. Part V (Passeriformes). Bull Fla State Mus Biol Sci 23:139–228

Brown JW, Rest JS, Garcia-Moreno J, Sorenson MD, Mindell DP (2008) Strong mitochondrial DNA support for a Cretaceous origin of modern avian lineages. BMC Biol 6:6

Carvalho de Taranto R, Loguercio MF, Bergqvist LP, Rocha-Barbosa O (2011) Comparative analysis of the hindlimb morphology of diogenornis fragilis (aves, ratites—paleocene) and flightless extant and extinct birds. IV Cong Lat Pal Vert

Cenizo MM, Agnolín FL (2010) The southernmost records of Anhingidae and a new basal species of Anatidae (Aves) from the lower–middle Miocene of Patagonia, Argentina. Alcheringa 34:1–22

Cheneval J (1983) Révision du genre Palaelodus Milne-Edwards, 1863 (Aves, Phoenicopteriformes) du gisement aquitanien de Saint-Gérand-le-Puy (Allier, France). Geobios 16:179–191

Cheneval J, Escuillié F (1992) New data concerning *Palaelodus ambiguus* (Aves: Phoenicopteriformes: Palaelodidae): ecological and evolutionary interpretations. In: Campbell KE Jr (ed) Papers in avian paleontology honoring Pierce Brodkorb. Los Angeles,

Clarke JA, Tambussi CP, Noriega JI, Erickson GM, Ketcham RA (2005) Definitive fossil evidence for the extant avian radiation in the Cretaceous. Nature 433:305–308

Cracraft J (1970) A new species of *Telmabates* (Phoenicopteriformes) from the lower Eocene of Patagonia. The Condor 72:479–480

Cracraft J (1971) Systematics and evolution of the Gruiformes (Class Aves) 2. Additional comments on the Bathornithidae, with descriptions of new species. Am Mus Novit 2449:1–14

De la Fuente M (2005) *Chelonoidis santafecina* Agnolín, 2004 sinónimo objetivo de *Testudo praestans* Rovereto, 1914. Ameghiniana 42:510–511

Degrange FJ (2012) Morfología del cráneo y complejo apendicular posterior de aves fororracoideas: implicancias en la dieta y modo de vida. Universidad Nacional de La Plata, Dissertation

Degrange FJ; Tambussi CP, Scaglia F, Dondas A, Taglioretti ML (2011) Hallazgo de un esqueleto completo y articulado de un nuevo Phorusrhacidae (Aves) en el Plioceno tardío de la Argentina. Actas IV Cong Lat Pal Vert

Degrange F, Tambussi C, Iglesias A, Zamuner A, Wilf P (2006) Primer registro de Aves para el Daniano. Ameghiniana 43:13R

Ericson PGP (2000) Systematic revision, skeletal anatomy, and paleoecology of the New World early Tertiary Presbyornithidae (Aves: Anseriformes). Paleobios 20:1–23

Ericson PGP, Anderson CL, Britton T, Elzanowski A, Johansson US, Källersjö M, Ohlson JI, Parsons TJ, Zuccon D, Mayr G (2006) Diversification of Neoaves: integration of molecular sequence data and fossils. Biol Lett 2:543–547

Feduccia A (1999) 1,2,3–2,3,4: Accommodating the cladogram. PNAS 96:4740–4742

Fountaine TM, Benton MJ, Dyke GJ, Nudds RL (2005) The quality of the fossil record of Mesozoic birds. Proc Biol Sci 7(272):289–294

Howard H (1955) A new wading bird from the Eocene of Patagonia. Am Mus Novit 1710:1–25

Iglesias A, Wilf P, Johnson K, Zamuner A, Cuneo NR, Matheos S (2007) A Paleocene lowland macroflora from Patagonia reveals significantly greater richness than North American analogs. Geology 35:947–950

Iglesias A, Artabe AE, Morel EM (2011) The evolution of Patagonian climate and vegetation, from the Mesozoic to the present. Biol J Linn Soc 103:409–422

Kurochkin EN, Dyke GJ (2010) A large collection of *Presbyornis* (Aves, Anseriformes, Presbyornithidae) from the late Paleocene and early Eocene of Mongolia. Geol J 45:375–387

Kurochkin EN, Dyke GJ, Karhu AA (2002) A new presbyornithid bird (Aves, Anseriformes) from the Late Cretaceous of southern Mongolia. Am Mus Novit 3386:1–11

Lindow BEK, Dyke GJ (2007) A small galliform bird from the Lower Eocene Fur Formation. Bull Geol Soc Den 55:59–63

Loomis F (1914) The Deseado formation of Patagonia. The Rumford press, Concord

Matthews SC (1973) Notes on open nomenclature and on synonymy list. Paleontology 16:713–719

Mayr G (2004) A partial skeleton of a new fossil loon (Aves, Gaviiformes) from the early Oligocene of Germany with preserved stomach content. J Ornithol 145:281–286

Mayr G (2009) Paleogene fossil birds. Springer-Verlag, Berlin Heidelberg

Mayr G, Alvarenga HMF, Clarke J (2011a) An *Elaphrocnemus*-like landbird and other avian remains from the late Paleocene of Brazil. Acta Palaeontol Pol 56:679–684

Mayr G, Alvarenga HMF, Mourer Chauviré C (2011b) Out of Africa: Fossil shed light on the origin of the hoatzin, an iconic neotropic bird. Naturwiss. doi:10.1007/s00114-011-0849-1

Mones A (1989) *Nomen Dubium* vs Nomen Vanum. J Vert Pal 9:232–234

Moreno FP, Mercerat A (1891) Catálogo de los pájaros fósiles de la República Argentina conservados en el Museo de La Plata. An Mus La Plata 1:7–71

Mourer Chauviré C (2000) A new species of Ameripodius (Aves: Galliformes: Quercymegapodiidae) from the Lower Miocene of France. Palaeontology 43:481–493

Murray PF, Vickers-Rich P (2004) Magnificent Mihirungs: the colossal flightless birds of the Australian dreamtime. Indiana University Press, Indiana

Noriega JI, Areta JI, Vizcaíno SF, Bargo MS (2011) Phylogeny and taxonomy of the Patagonian Miocene Falcon Thegornis musculosus Ameghino, 1895 (Aves: Falconidae). J Paleontol 85:1089–1104

Olson S (1985) The fossil record of birds. In: Farner D, King J, Parkes K (eds) Avian biology. Academic Press, New York

Olson SL, Alvarenga HMF (2002) A new genus of small teratorn from the Middle Tertiary of the Taubaté Basin, Brazil (Aves: Teratornithidae). Proc Biol Soc Wash 115:701–705

Olson S, Feduccia A (1980) *Presbyornis* and the origin of the Anseriformes (Aves: Charadriomorphae). Smith Cont Zool 323:1–24

Patterson B (1941) A new phororhacoid bird from the Deseado formation of Patagonia. Geol Ser Field Mus Nat Hist 8:49–54

Petrulevicius J, Tambuscci CP (1995) Primer hallazgo de una pluma para el Paleógeno de América del Sur. Acta Geol Lill 18:176–177

Sallaberry MA, Yury-Yáñez RE, Otero RA, Soto-Acuña S, Torres T (2010) Eocene birds from the western margin of southernmost South America. J Paleontol 84:1061–1070

Smith VS, Ford T, Johnson K, Johnson PCD, Yoshizawa K, Light JE (2011) Multiple lineages of lice pass through the K-Pg boundary. Biol Lett 7:782–785

Tambussi CP (1989) Las aves del Plioceno tardío-Plesitoceno temprano de la Provincia de Buenos Aires. Universidad Nacional de La Plata, Dissertation

Tambussi CP (1995) The fossil Rheiformes from Argentina. Cour Forch Senck 181:121–129

Tambussi CP (2011) Paleoenvironmental and faunal inferences based upon the avian fossil record of Patagonia and Pampa: what works and what does not. Biol J Linn Soc 103:458–474

Tambussi CP, Noriega JI (1996) Summary of the Avian Fossil Record from Southern South America. In: Arratia G (ed) Contributions of the southern south America to vertebrate paleontology. Müncher Geowissenschaftliche Abhandlungen

Tambussi CP, Noriega JI (1998) Registro de Presbiornítidos (Aves, Anseriformes) en sedimentitas de la Formación Vaca Mahuida (La Pampa, Argentina). Asociación Paleontológica Argentina pub. esp. 5:51–54.

Tonni EP (1980) The present state of knowledge of the Cenozoic birds of Argentina. Contrib Sci 330:104–114

Tonni EP, Tambussi CP (1986) Las aves del Cenozoico de la República Argentina. Actas IV Cong Arg Pal Bioestrat 2:131–142

van Tuinem M, Stidham TA, Hadly EA (2006) Tempo and mode of modern bird evolution observed with large-scale taxonomic sampling. Hist Biol 18:205–221

Worthy T (2008) Tertiary fossil waterfowl (Aves: Anseriformes) of Australia and New Zealand. Dissertation, University of Adelaida.

Yury-Yáñez RE, Otero RA, Soto-Acuña S, Suárez ME, Rubilar-Rogers D, Sallaberry M (2012) First bird remains from the Eocene of Algarrobo, central Chile. And Geol 39 (3): 548-557

Chapter 6
Eocene Birds from Antarctica and Their Relationships with Those of South America

All avian fossils from Antarctica (e.g. Chatterjee 2002; Chatterjee et al. 2006; Clarke et al. 2005, 2006; Coria et al. 2007; Jadwiszczak 2011; Tambussi and Acosta Hospitaleche 2007; Tambussi and Tonni 1988; Tambussi et al. 1994, 1995, 2005, 2006; Tambussi and Degrange 2012; Tonni and Tambussi 1985) appear to represent members of the anatomically modern clade Neornithes. This is true not only for the Cenozoic but also for the Cretaceous record that comprises more archaic taxa such as enantiornithines, hesperornithiformes, or basal ornithurines. Eocene deposits of La Meseta Formation have yielded considerable numbers of bird remains, including those of seabirds and continental ones.

Seabirds are an ecologically important group characterized by their dependence on the marine environment. Antarctica hosts six breeding species of penguins (Sphenisciformes) today. They have a near-shore aquatic lifestyle, gregarious habits, non-pneumatic bones, and wings transformed into flippers. Yet, the fossil record indicates that a highly diverse array of now-extinct taxa once inhabited Antarctic coastlines. Penguins constitute the most frequent fossil remains from the James Ross Basin, especially from the uppermost unit of La Meseta Formation (Submeseta Allomember or Telm7, Priabonian, Late Eocene, ~ 34–37 Ma). Starting at the Late Paleocene (Tambussi et al. 2005), its record ends at the Eocene including 15 species, 9 of which would have coexisted (Myrcha et al. 2002; Tambussi et al. 2006; Tambussi and Acosta Hospitaleche 2007). However, Jadwiszczak and Thomas (2011) synonymized four species previously recognized. Nevertheless, diversity—and frequency of the remains—is still astounding. High diversity could have resulted in an ecological segregation of penguins related to differences in breeding chronology, foraging behavior, or life history tactics. It is not new that this type of ecological segregation occurs in current penguin colonies (Bunnefeld et al. 2011) but in no case, is diversity as dramatic as in the Eocene of Seymour. Three localities in Magallanes, southern Chile, whose stratigraphic context indicates a positive correspondence with the geological units of Seymour Island, have yielded two different groups of penguins. The first group is similar in size to the smallest taxa previously

C. P. Tambussi and F. J. Degrange, *South American and Antarctic Continental Cenozoic Birds*, SpringerBriefs in Earth System Sciences, DOI: 10.1007/978-94-007-5467-6_6, © The Author(s) 2013

described from Seymour Island (e.g., *Marambiornis* Myrcha et al. 2002, *Mesetaornis* Myrcha et al. 2002, and *Delphinornis* Wiman 1905) and the second is similar in size to the bigger taxa. Moreover, Chile and Seymour share records of *Palaeeudyptes* Huxley 1859, one of the most widespread penguin genera (including also New Zealand, Chile, and Perú) in the southern hemisphere during the Eocene (Sallaberry et al. 2010). Indeed, this is not as striking as it may seem. Living penguins generally do not migrate great distances, but Adeliae penguin for example migrates about 600 km north of the Antarctic continent (Dunn et al. 2011). The current distance between the southernmost end of South America and Antarctica is about 1,000 km and it is known that this gap was smaller during the Eocene–Oligocene times (Fig. 2.3). It is very reasonable to think that an exchange—or dispersion—of fauna have been possible in this marine scenario. Apparently, Antarctic penguins had a complex history involving multiple dispersal and extinction events. Is not a goal of this work to devote to these issues in-depth? We refer readers who are interested in these topics to learn more about the diversity and evolution of Antarctic penguins in Jadwiszczak (2009, 2010 and the literature cited therein), Ksepka et al. (2006), Slack et al. (2006), Tambussi et al. (2005, 2006) and Tambussi and Acosta Hospitaleche (2007).

A broad picture of the Eocene continental avifaunas of West Antarctica has emerged in the past years, but the increase of our knowledge is low. The mid-Tertiary lacustrine sediments of King George Island (the largest of the South Shetland Islands) preserved four types of footprints which belong to the tetra-dactyle footprint *Antarctichnus fuenzalidae*, shorebirds, non-volant ground birds that could belong to either ratites or gruiforms, and probably an anatid (Covacevich and Lamperein 1972; Covacevich and Rich 1982). The ichnofossils include both solitary and group activities with their hypothetical avian tracemakers (Tambussi and Acosta Hospitaleche 2007). Footprints have also been reported belonging to a bird with three anterior long toes that could correspond to Ratitae or Phorusrhacidae (Case et al. 1987).

Within the non-penguin materials from Seymour Island housed at Museo de La Plata, there is also a near-complete left coracoid (Fig. 6.1a) collected in the Cucullaea I Allomember (Telm 5, Early-Middle Eocene, Ypresian/Lutetian, ~49–52 Ma) which can be assigned to a loon (Gaviiformes) with some degree of reliability (Tambussi et al. 2012). The coracoid has a short and robust shaft; the cotyla scapularis is subtriangular and deep; the facies articularis humeralis is flat, oval, and broad; the procoracoid process is broken but the base is very broad; the processus acrocoracoideus is partially broken but it was very well developed; the foramen n. supracoracoidei is incospicuous; the facies articularis sternalis is broad at the level of the angulus medialis; the impressio m. stercoracoidei is shallow and the sulcus m. supracoracoidei is broad and deep; the impressio lig. acrocoraco-humeralis is conspicuous, deep, and situated proximal to the facies articularis humeralis. The Antarctic coracoid is smaller in size with the living *Gavia immer*. Although this fossil cannot be distinguished from the living taxa, the morphology of the only available specimen is insufficient to determine the specific level at the moment. A deep analysis of this specimen is now in progress. Extant loons (four species of the genus *Gavia*) are foot propelled divers found in North America and northern Eurasia. They breed at northern freshwater sites, but winter along sea coasts in temperate areas (Carboneras 1992). Loons had a more southerly distribution than the present day, and their fossils have been found in California, Florida, Italy, Austria, Chile, and Antarctica (Chatterjee 2002; Mayr 2004; Mlikovsky 1998; Olson 1985, 1992). The earliest fossil gaviiform (Lambrecht 1929; Olson 1992) that resembles the highly derived bone of modern loons (Mayr 2004) had been described from the Upper Cretaceous of Chile (Quinriquina Formation) and Antarctica (López de Bertodano Formation). It is likely that both records belong to the same species, *Neogaeornis wetzeli* Lambrecht 1929. The presence of a loon in the Eocene of La Meseta Formation constitutes the youngest record of gaviids in the southern hemisphere and also extends the permanence of this Holartic lineage to the Eocene in the southern hemisphere (Tambussi et al. 2012).

A number of bird bones excavated from the Eocene deposits of La Meseta Formation were attributed to pelagornithid birds (Tambussi and Acosta Hospitaleche 2007; Tonni 1980; Tonni and Tambussi 1985) (Fig. 6.1b, c). The Pelagornithidae, commonly called pelagornithids or pseudodontorns, are a bony-toothed extinct family of large seabirds. Their fossil remains have been found all over the world (England, Europe, North America, Japan, New Zealand, Africa, Chile, and Perú, Harrison and Walker 1976; McKee 1985; Mayr and Rubilar-Rogers 2010; Mayr 2011; Olson 1985; Walsh and Hume 2001; Warheit 1992), in rocks dating between the Late Paleocene and the Pliocene–Pleistocene boundary, in all sorts of climates. They were the dominant seabirds of most oceans throughout most of the Cenozoic. Their most notable trait is the presence of tooth-like points on the edge of the premaxillary and mandibular bones that contained Volkmann's canals (the transverse channels that are absent in true teeth and that interconnect the Haversian canals of the tissue). Pelagornithids comprise medium species with the size of albatrosses and very gigantic taxa up to 6 m of wingspans (e.g., they are among the largest flying birds ever lived, Mayr and Rubilar-Rogers 2010). They range from generalists that could probably undertake flapping flight to highly specialized gliders. Antarctic pelagornithids seem to fit into these two morphotypes. Interestingly, pelagornithids of Seymour Island are found in the company of penguins while those of the northern hemisphere are associated with the wing-propelled diver plotopterids (Ploptopteridae). Warheit (1992) has suggested that a worldwide Late Eocene oceanic cooling could be the cause for that association.

Procellariiformes include the modern albatrosses, petrels, and storm-petrels. Modern albatrosses (Diomedeidae) are worldwide pelagic and gliding seabirds. However, its fossil record is fairly from the northern hemisphere, where they appear since the Late Oligocene (Tambussi and Tonni 1988; Mayr 2009). A weathered tarsometatarsus from the La Meseta Formation at Seymour Island (Noriega and Tambussi 1996; Tambussi and Tonni 1988) can be unambiguously assigned to this family. Additional fossil specimens housed at Museo de La Plata (Fig. 6.1e) could be also assigned to Procellariidae (Noriega and Tambussi 1996 and this paper).

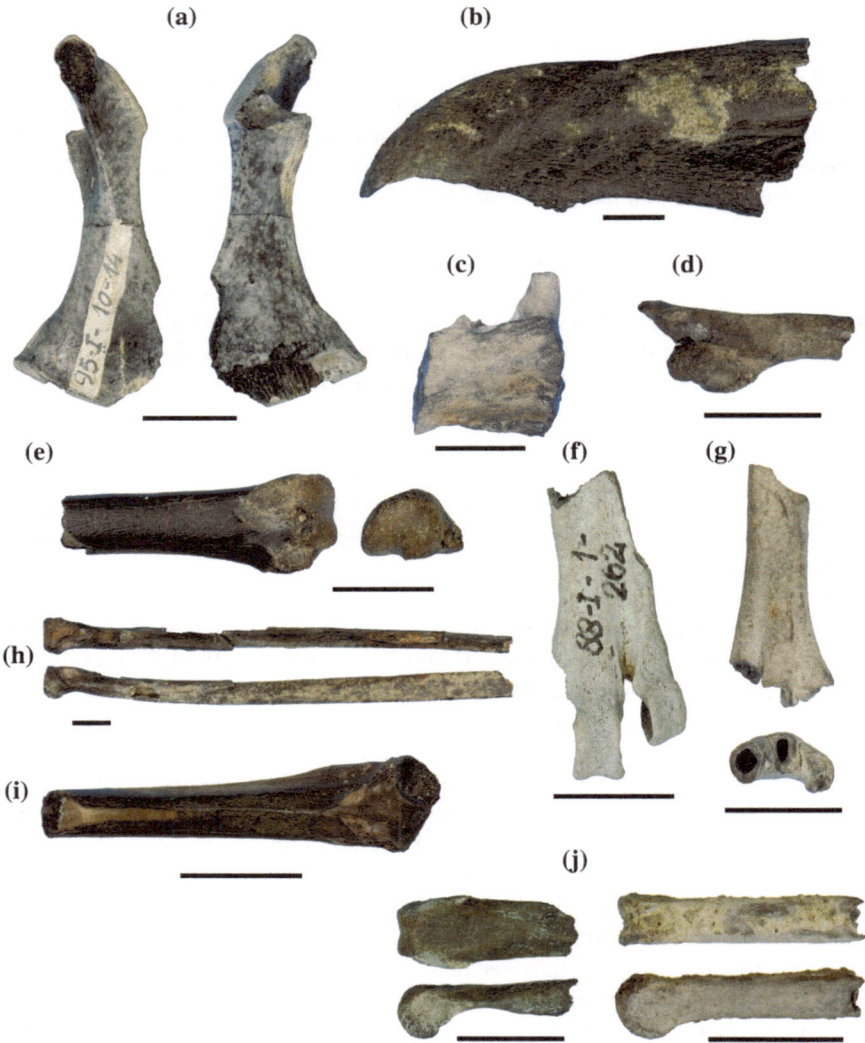

Fig. 6.1 Fossil birds from Antarctica: **a** Gaviiformes, left coracoid (MLP 95-I-10-14); Pelagornithidae, **b** distal fragment of a beak (MLP 08-XI-30-42), **c** fragment of mandible (MLP 83-V-30-2), **d** Presbyornithidae, left scapula (MLP 96-I-5-19), **e** Procellariidae, right ulna (MLP 91-II-4-6), **f** Charadriiformes, distal fragment (MLP 88-I-1-262), **g** Gruiformes, right distal tarsometatarsus (MLP 90-I-20-9), **h** hoenicopteriformes, right radius (MLP 87-II-1-2), **i** Charadriidae, left proximal ulna (MLP 95-I-10-9), **j** Falconiformes, phalanges (MLP 92-II-2-7). Materials **a**, **d**, **e**, **i** and **j** were recovered from Cucullaea I Allomember of the marine La Meseta Formation at Seymour Island (Telm 5, Early-Middle Eocene, Ypresian/Lutetian, ~49–52 Ma)

Charadriiforms, shorebirds, and waders are a heterogeneous and polymorphic group of birds of small to moderate size. Because of its high diversity, it is difficult to diagnose the group with morphological characters. They frequent open inland and coastal/marine wetlands. Charadriiforms (Fig. 6.1f, i) from La Meseta Formation are represented by a right scapula, distal left tarsometatarsus, and proximal left ulna. All bones are isolated and represent more than one individual.

A distal fragment of a right tarsometatarsus probably belonging to Gruiforms was found at the upper level of La Meseta Formation (Fig. 6.1g). Trochleas are completely absent but the preserved part reveals that the bases of trochlear II and III are very close, separated from trochlea IV by a conspicuous groove. The size and shape fits well with the living American coot. Unfortunately, the material is not preserved enough to allow a more accurate identification.

A notoriously long, slender, and slightly curved incomplete right radius (Fig. 6.1h) of a flamingo was reported by Noriega and Tambussi (1996). Our revision confirms the assignation. Flamingos (Phoenicopteridae) are gregarious, long-legged filter-feeders, and invariably associated with warm temperatures, brackish or saltwater lakes and lagoons. The fossil record of Phoenicopteriformes goes back into the Eocene of North American and Messel deposits (*Juncitarsus*, Olson and Feduccia 1980; Peters 1987; Mayr 2009). Paleolodids (Phoenicopteriformes), to which flamingos are most closely related, are now known from the Early Miocene of New Zealand (Worthy et al. 2010). The Antarctic material is too incomplete to assess its exact position, but it is the earliest record of a Phoenicopteriformes in the southern hemisphere.

Tambussi et al. (1995) have reported the presence of a diurnal bird of prey, Falconiformes Polyborinae, at La Meseta Formation. The material constitutes the oldest record for the Falconidae family and consists of a tarsometatarsus (not figured here) whose morphology resembles that of living polyborines in having the trochlea for the second digit shorter and wider than the trochlea for the digit four, bearing a planar projection. Living Polyborines are vulture-like falconids with scavenging habits that occur exclusively in the Americas, mainly in the Neotropical regions. The animal would have reached a body mass of about one kilogram and the size of the living caracara *Polyborus plancus* (Tambussi and Acosta Hospitaleche 2007). Additionally, Falconiformes are also represented by a pedal phalanx (Fig. 6.1j) figured here for the first time. Interestingly, nowadays, Falconidae is a family of worldwide distribution (except Antarctica and the Arctic), with the greatest diversity concentrated mainly in South America (Olson 1976; White et al. 1994).

From the topmost levels of the Submeseta Allomember, part of the near-shore deposits of the La Meseta Formation, likely Late Eocene (ca, 36 Ma Dutton et al. 2002; Reguero et al. 2002), two different taxa of large flightless birds from Antarctica have been described: a ratite (Tambussi et al. 1994) and a supposed phorusrhacid (Case et al. 1987, 2006). Strictly, Late Eocene terrestrial birds of Antarctica raise some interesting biogeographic issues that we will discuss below.

Antarctic phorusrhacid records require special considerations. New perspectives on the materials reported by Case et al. (1987, 2006) who were followed by other authors (Hospitaleche Tambussi and Acosta 2007), indicate that their systematic locations were wrong. The fragment of a beak (cast UCR 22175) presents lateral grooves that are not present in any known phorusrhacid (Degrange 2012). It is noteworthy that this feature resembles Pelecaniformes Pelagornithidae. However, at least at the moment this assignment cannot be corroborated. The other remain, a distal end of a tarsometatarsus (cast UCR 22175) presents the trochlea metatarsi II proximally retracted and twisted caudally, a feature that does not correspond to any known Phorusrhacidae (Degrange 2012; Cenizo 2012).

Regarding the ratite, the Antarctic material is a distal tarsometatarsus with a "large, narrow trochlea for digit III, which is projected moderately beyond the trochlea for digit II with straightened margins bordering a deep groove. Trochlea II has a wide articular surface and extends posteriorly more than trochlea III. The lateral margin of trochlea III allow us to infer that the intertrochlear space between trochlea III and IV extends proximately beyond trochleae II and III" (Tambussi et al. 1994, p. 606).

More recently, Case et al. (2006) reported the presence of a cursorial bird from the Late Cretaceous (Maastrichtian) of Vega Island, Antarctic Peninsula and would correspond in an ancestral form to the Cariamidae-Phorusrhacidae or to a basal Cariamidae, according to the authors. The affinities of these remains could not be corroborated. However, at least one specimen shown by Case and colleagues during 2006 SVP meeting (USA) is a tibiotarsus belonging to Sphenisciformes.

Most of the ratites (ostriches, emu, cassowaries, forest-dwelling kiwis, and rheas) live currently in the southern hemisphere, and all of them lack a keel on the sternum, a character associated with flightlessness. The estimated body mass of the Antarctic specimen is approximately 60 kg (Vizcaíno et al. 1998), greater than that of the Greater rhea (23–25 kg according to Picasso 2010) but considerably lower than that of the adult male ostrich (90 kg sensu Alexander 1985). Although the phylogenetic position of this animal is far from being elucidated, some biogeographic considerations can be made.

In the context of what has been found, the fossil is significant in that it is from a land-dwelling bird. As was previously mentioned, many of the fossil birds found in Antarctica were birds which lived along the shoreline, and the ratite represents the oldest strictly terrestrial bird found in Antarctica. For now, we will assume that the bird truly does represent a previously unknown branch of a group which primarily resided in the southern continents.

The presence of a strictly ground bird strongly supports the idea that West Antarctica was used as dispersal route for obligate terrestrial organisms at least during the earliest Paleogene, when the opening of the Drake Passage began. Beyond the phylogenetic relationships of this animal with the other ratites, this is a case of southerly trans-Pacific disjunctions that are among the most conspicuous and notorious of all known distributional patterns between biogeographers (Brundin 1966; Cracraft 2001).

All these birds were accompanied by a rich mammal fauna found in the middle levels of La Meseta Formation composed of small marsupials and Microbiotheria Didelphimorphia (Goin and Carlini 1995; Goin et al. 1999, 2006), some

Phyllophaga xenarthrans (the earliest record of this group), and astrapoterids, and litopterns (Bond et al. 1990; Hooker 1992; Marenssi et al. 1994; Vizcaíno et al. 1997; Bond et al. 2006), and gondwanatherians Sudamericidae (Reguero et al. 2002; Goin et al. 2006).

The reconstruction of this fauna and its population structure suggest a coastal environment, with *Nothofagus* forests that developed close to a volcanic mountain range in warm, humid climates (Reguero et al. 1998, 2002). The floristic association of the Early-Middle Eocene of the Seymour Island suggests the development of a mixed mesophytic forest, which denotes a warm, to cold environment with marked seasonality, and average minimum temperature of about −3 °C (Chornogubsky 2010).

References

Alexander RMc (1985) The Legs of Ostriches (*Struthio*) and Moas (*Pachyornis*). Acta Biotheor 34:2–4

Bond M, Reguero MA, Vizcaíno SF, Marenssi S (1990) A new south American ungulate (mammalian Litopterna) from the Eocene of the Antarctic Peninsula. In: Francis JE, Pirrie D, Alistair Crame J (eds) Cretaceous-tertiary high-latitude palaeoenvironments: James Ross Basin, Antarctica. Geological Society of London, London

Bond M, Reguero MA, Vizcaíno SF, Marensi SA (2006) A new "South American ungulate" (Mammalia: Litopterna) from the Eocene of the Antarctic Peninsula Geological Society of London Special Publications, pp 163–176

Brundin L (1966) Transantarctic relationships and their significance as evidenced by chironomid midges: with a monograph of the sub-families Podonominae and Aphroteniinae and the Austral Heptagyiae. K Svenska Vet Hand 11:1–472

Bunnefeld N, Börger L, van Moorter B, Rolandsen CM, Dettki H, Solberg EJ, Ericsson G (2011) A model-driven approach to quantify migration patterns: individual, regional and yearly differences. J Anim Ecol 80:466–476

Carboneras C (1992) Family Gaviidae (Divers). In: del Hoyo J, Elliott A, Sargatal J (eds) Handbook of the birds of the world. Lynx Edicions, Barcelona

Case J, Woodbourne M, Chaney D (1987) A gigantic Phororhacoid (?) bird from Antarctica. J Pal 61:1280–1284

Case J, Reguero M, Martin J, Cordes-Person A (2006) A cursorial bird from the Maastrichtian of Antarctica. J Vert Pal 26:48A

Chatterjee S (2002) The morphology and systematics of *Polarornis*, a Cretaceous loon (Aves: Gaviidae) from Antarctica. In: Zhou Z, Zhang F (eds) Proceedings of the 5th symposium of the society of avian paleontology and evolution, Beijing

Cenizo M (2012) Review of the putative Phorusrhacidae from the Cretaceous and Paleogene of Antarctica: new records of ratites and pelagornithid birds. Polish Polar Research 33:239–258

Chatterjee S, Martinioni D, Novas F, Mussel F, Templin R (2006) A new fossil loon from the late Cretaceous of Antarctica and early radiation of foot-propelled diving birds. J Ver Pal 26:49A

Chornogubsky L (2010) Los polidolópidos de la Isla Marambio (Península Antártica) y Paso del Sapo (Provincia del Chubut, Argentina): ajuste a la Regla de Bergmann. Ameghiniana 47: 7R

Clarke JA, Tambussi CP, Noriega JI, Erickson GM, Ketcham RA (2005) Definitive fossil evidence for the extant avian radiation in the Cretaceous. Nature 433:305–308

Clarke JA, Tambussi CP, Noriega JI, Erickson GM, Ketcham RA (2006) Corrigendum to Definitive fossil evidence for the extant avian radiation in the Cretaceous. Nature 444:780

Covacevich V, Lamperein C (1972) Hallazgos de icnitas en Península Fildes, Isla Rey Jorge, Archipiélago Shetland del Sur, Antártica. Ser Cient INACH (Inst Ant Chi) 1:55–74

Coria RA, Tambussi CP, Moly JJ, Santillana S, Reguero M (2007) Nuevos restos de dinosauria del Cretácico de las islas James Ross y Seymour, Península Antártica. Jornadas de Comunicaciones sobre Investigaciones Antárticas. Ciudad Autónoma de Buenos Aires, septiembre 2007, Resumen expandido

Covacevich V, Rich PV (1982) New birds ichnites from Fildes Peninsula, King George Island, West Antarctica. Int Univ Geol Sci Ser B 4:245–254

Cracraft J (2001) Avian evolution, Gondwana biogeography and the Cretaceous-Tertiary mass extinction event. Proc Royal Soc London 268:459–469

Degrange FJ (2012) Morfología del cráneo y complejo apendicular posterior de aves fororracoideas: implicancias en la dieta y modo de vida. Universidad Nacional de La Plata, Dissertation

Dunn MJ, Silk JRD, Trathan PN (2011) Post-breeding dispersal of Adelie penguins (Pygoscelis adeliae) nesting at Signy Island, South Orkney Islands. Polar Biol 34:205–214

Dutton AL, Lohmann KC, Zinsmeister WJ (2002) Stable isotope and minor element proxies for Eocene climate of Seymour Island. Paleocean, Antarctica. doi:10.1029/2000PA000593

Goin FJ, Carlini AA (1995) An early Tertiary microbiotheriid marsupial from Antartica. J Vert Pal 15:205–207

Goin FJ, Case JA, Woodburne MO, Vizcaíno SF, Reguero MA (1999) New discoveries of "opposum-like" marsupials from Antarctica (Seymour Island, medial Eocene). J Mammal Evol 6:335–365

Goin FJ, Pascual R, Koenigswald WV, Woodburne MO, Case JA, Reguero M, Vizcaíno SF (2006) First Gondwanatherian Mammal from Antarctica. In: Francis JE, Pirrie D, Crame JA (eds) Cretaceous-tertiary high-latitude paleoenvironments, James Ross Basin, Antarctica. The Geological Society, Special Publication 258, London

Harrison JO, Walker CA (1976) A review of the bony-toothed birds (Odontopterygiformes): with descriptions of some new species. Tert Res Special Pap 2:1–62

Hooker JJ (1992) An additional record of a placental mammal (order Astrapotheria) from the Eocene of west Antarctica. Antarct Sci 4:107–108

Jadwiszczak P (2009) Penguin past: the current state of knowledge. Pol Polar Res 30:3–28

Jadwiszczak P (2010) New data on the appendicular skeleton and diversity of Eocene Antarctic penguins. In: Nowakowski D (ed) Morphology and systematics of fossil vertebrates. DN Publisher, Poland

Jadwiszczak P (2011) New data on morphology of late Eocene penguins and implications for their geographic distribution. Ant Sci 23:605–606

Jadwiszczak P, Mörs T (2011) Aspects of diversity in early Antarctic penguins. Acta Palaeontologica Polonica 56:269–27

Ksepka DT, Bertelli S, Giannini NP (2006) The phylogeny of the living and fossil Sphenisciformes (penguins). Cladistics 22:412–441

Lambrecht K (1929) Mesozoische und tertiäre Vogelreste aus Siebenbürgen. In: Csiki E (ed) Xe congrés international de zoologie. Stephaneum, Budapest

Marenssi SA, Reguero MA, Santillana SA, Vizcaíno SF (1994) Eocene land mammals from Seymour Island, Antarctic palaeobiogeographical implications. Antarct Sci 6:3–15

Mayr G (2004) A partial skeleton of a new fossil loon (Aves, Gaviiformes) from the early Oligocene of Germany with preserved stomach content. J Ornithol 145:281–286

Mayr G (2009) Paleogene fossil birds. Springer, Berlin

Mayr G (2011) Two-phase extinction of "Southern Hemispheric" birds in the Cenozoic of Europe and the origin of the Neotropic avifauna. Palaeobio Palaeoenv 91:325–333

Mayr G, Rubilar-Rogers D (2010) Osteology of a new giant bony-toothed bird from the Miocene of Chile, with a revision of the taxonomy of Neogene Pelagornithidae. J Vert Pal 30:1313–1330

Myrcha A, Jadwiszczak P, Tambussi CP, Noriega JI, Gaździcki A, Tatur A, del Valle R (2002) Taxonomic revision of Eocene Antarctic penguins based on tarsometatarsal morphology. Polish Polar Research 23:5–46

McKee JWA (1985) A pseudodontorn (Pelecaniformes: Pelagornithidae) from the middle Pliocene of Hawera, Taranaki, New Zealand. New Zeal J Zool 12:181–184

Mlikovsky J (1998) A new loon (Aves: Gaviidae) from the middle Miocene of Austria. Ann Natur Mus Wien 99:331–339

Olson SL (1976) Oligocene fossils bearing on the origins of the Todidae and the Momotidae (Aves: Coraciiformes). Smithson Contrib Paleobiol 27:111–119

Olson S (1985) The fossil record of birds. In: Farner D, King J, Parkes K (eds) Avian biology. Academic Press, New York

Olson SL (1992) *Neogaeornis wetzeli* Lambrecht, a Cretaceous loon from Chile (Aves: Gaviidae). J Vert Pal 12:122–124

Olson S, Feduccia A (1980) *Presbyornis* and the origin of the Anseriformes (Aves: Charadriomorphae). Smith Cont Zool 323:1–24

Peters DS (1987) Ein "Phorusrhacidae" aus dem Mittel-Eozan von Messel (Aves, Gruiformes, Cariamae). Doc Lab Géol Lyon 99:71–87

Picasso MBJ (2010) Crecimiento y desarrollo de los componentes musculares y óseos asociados a la locomoción durante la vida postnatal de Rhea americana (Aves: Palaeognathae). Dissertation, Universidad Nacional de La Plata

Reguero MA, Vizcaíno SF, Goin FJ, Marenssi SA, Santillana SN (1998) Eocene high-latitude terrestrial vertebrates from Antarctica as biogeographic evidence. Asoc Pal Arg 5:185–198

Reguero MA, Marenssi SA, Santillana SN (2002) Antarctic Peninsula and Patagonia Paleogene terrestrial environments: biotic and biogeographic relationships. Palaeogeog Palaeoecol 2776:1–22

Sallaberry MA, Yury-Yáñez RE, Otero RA, Soto-Acuña S, Torres T (2010) Eocene birds from the western margin of southernmost South America. J Paleontol 84:1061–1070

Slack KE, Jones CM, Ando T, Harrison GL, Fordyce RE, Arnason U, Penny D (2006) Early penguin fossils, plus mitochondrial genomes, calibrate avian evolution. Mol Biol Evol 23:1144–1155

Tambussi CP, Acosta Hospitaleche C (2007) Antarctic birds (Neornithes) during the Cretaceous-Eocene times. Rev Asoc Geol Arg 62:604–617

Tambussi CP, Noriega JI (1996) Summary of the avian fossil record from Southern South America. In: Arratia G (ed) Contributions of the southern south America to vertebrate paleontology. Müncher Geowissenschaftliche Abhandlungen, München

Tambussi CP, Tonni EP (1988) Un Diomedeidae (Aves: Procellariiformes) del Eoceno tardío de la Antártida. In: Quiroga J, Cione A (eds) V Jor Arg Pal Vert, pp 34–35

Tambussi CP, Noriega JI, Gazdzicki A, Tatur A, Reguero MA, Vizcaíno SF (1994) Ratite bird from the Paleogene La Meseta formation, Seymour Island, Antarctica. Pol Polar Res 15:15–20

Tambussi CP, Noriega JI, Santillana S, Marenssi S (1995) Falconid bird from the Middle Eocene La Meseta formation, Seymour Island. West Antarct J Vert Pal 15:55A

Tambussi CP, Reguero MA, Marenssi SA, Santillana SN (2005) *Crossvallia unienwillia*, a new Spheniscidae (Sphenisciformes, Aves) from the Late Paleocene of Antarctica. Geobios 38:667–675

Tambussi CP, Acosta Hospitaleche C, Reguero MA, Marenssi SA (2006) Late Eocene penguins from West Antarctica: systematics and biostratigraphy. In: Francis JE, Pirrie D, Crame JA (eds) Cretaceous-Tertiary high-latitude palaeo environments, James Ross Basin, Antarctica. Geological Society of London, London

Tambussi CP, Degrange FJ, Reguero MA, Marenssi SA, Santillana SN (2012). Antarctic Eocene loon (Gaviiformes): last refuge of survivor of a long typically Holarctic lineage? In: SCAR open science conference (OSC), Portand. http://scar2012.geol.pdx.edu/themes.php. Accessed 1 May 2012

Tonni EP (1980) The present state of knowledge of the Cenozoic birds of Argentina. Contrib Sci 330:104–114

Tonni EP, Tambussi CP (1985) New remains of Odontopterygia (Aves: Pelecaniformes) from the early Tertiary of Antarctica. Ameghiniana 21:121–124

Vizcaíno SF, Bond M, Reguero MA, Pascual R (1997) The youngest record of fossil land
    mammals from Antarctica; its significance on the evolution of the terrestrial environment of
    the Antarctic Peninsula during the late Eocene. J Paleont 71:348–350
Vizcaíno SF, Reguero M, Goin F, Tambussi C, Noriega JI (1998) Community structure of Eocene
    terrestrial vertebrates from Antarctic Peninsula. Ameghiniana Pub Esp 5:177–183
Walsh SA, Hume JP (2001) A new neogene marine avian assemblage from north-central Chile.
    J Vert Pal 21:484–491
Warheit KI (1992) A review of the fossil seabirds from the Tertiary of the North Pacific: plate
    tectonics, paleoceanography, and faunal change. Paleobiol 18:401–424
White CM, Olsen PD, Kiff LF (1994) Family falconidae. In: del Hoyo J, Elliott A, Sargatal J
    (eds) Handbook of birds of the world, volume 2 (New World Vultures to Guineafowl). Lynx
    Edicions, Barcelona
Worthy TH, Tennyson AJD, Archer M, Scofield RP (2010) First record of *Palaelodus* (Aves:
    Phoenicopteriformes) from New Zealand. Rec Aust Mus 62:77–88

Chapter 7
Neogene Birds of South America

7.1 The Lower-Middle Miocene Santa Cruz Formation

The continental vertebrate collection of the Santa Cruz Formation (Late-early Miocene) is known worldwide by its abundance and diversity (Hatcher 1903; Tauber 1997a, b; Vizcaíno et al. 2006, 2010). With regard to the bird fossil record, the taxonomical and morphological diversity is also really high (Tonni 1980; Olson 1985; Tambussi and Noriega 1996; Alvarenga and Höfling 2003; Agnolín 2004, 2006a, b, 2007, 2009b; Noriega et al. 2009; Cenizo and Agnolín 2010; Tambussi 2011; Degrange et al. 2012). The initial collection was largely made by Carlos Ameghino by the end of the nineteenth century and the remains were studied and nominated by his brother, Florentino Ameghino (1887, 1889, 1891a, b, 1895, 1899) (Table 5.1). Subsequent fieldworks achieved by the Princeton University and the result was the addition of new bird remains illustrated and described by Sinclair and Farr (1932) in the Reports of the Princeton University Expeditions to Patagonia (Hatcher in charge). While several of the new specimens correspond to complete and beautifully preserved skeletons of *Psilopterinae phorusrhacids*, other remains are incomplete, isolated, broken, and eroded materials, and in some cases wrongly identified (e.g., Degrange 2012).

Only few species are represented by nearly complete skeletons (e.g., *Psilopterus lemoinei, Psilopterus bachmanni, Patagornis marshi*, and the falconid *Thegornis musculosus*, see below). Descriptions of the remains originally collected are bare, inaccurate situations that makes difficult the correct systematic location and founding of the phylogeneic relationship of most of the santacrucian taxa (Olson 1985; Tonni 1980; Tambussi and Noriega 1996); although, there have been several recent efforts to shed some light on this situation (Alvarenga and Höfling 2003; Agnolín 2004, 2006a, b, 2007, 2009b; Tambussi 2011, Tambussi and Degrange 2011; Degrange et al. 2012).

Till date, the bird fossil record of the Santa Cruz Formation includes at least 18 species located in 15 genera and 9 families (Degrange et al. 2012).

C. P. Tambussi and F. J. Degrange, *South American and Antarctic Continental Cenozoic Birds*,   59
SpringerBriefs in Earth System Sciences, DOI: 10.1007/978-94-007-5467-6_7,
© The Author(s) 2013

Opistodactylus patagonicus is without doubt a Rheiformes, the clade endemic of South America that includes rheas. In fact, the tibiotarssus and tarsometatarsus (Fig. 7.1a) on which Ameghino found this species does not differ much on its morphology and proportions from their extant relatives (Tambussi 1995). Both living species of the family Rheidae are flightless large fast-running birds. They are the greater living birds of the Americas.

The closest living relatives of Rheas are the tinamous (Order Tinamiformes) that also have the first record during the Miocene. Tinamids were shown to be monophyletic and their current distribution is restricted to Central and South America. Some species are related to arid environments (Nothurinae) and others to forest ("Tinaminae"). In sediments of the Santa Cruz Formation, at least three species of Tinamidae are registered. Three unnamed species are from Monte Observación (= Cerro Observatorio), Monte León, and Cañadón de las Vacas localities (Chiappe 1991; Bertelli and Chiappe 2005). These remains represent two morphotypes of Nothurinae, the aridland tinamous (Degrange et al. 2012). However, the state of preservation of the material collected (fragmentary coracoids, humerus and tibiotarsi) has not allowed a more precise identification. Recently a new fossil, tinamous *Crypturellus reai*, was described by Chandler (2012). The material, a complete left humerus, was collected by Barnum Brown at Cañadón de Las Vacas, at Santa Cruz Province (Argentina) during the Princeton expedition to Patagonia (1898–1899) (Chandler 2012). We could not study the material directly but judging by the images of Chandler's paper, we doubt that the systematic position of the fossil is appropriate. For example, the proximal extremity is similar to tinamids, whereas the distal one deeply resembles those of Cracids (e.g., *Ortalis*). Currently, species of *Crypturellus* are forest-dwelling birds, distributed and associated with enclosed environments far north from Santa Cruz.

Falconidae are a group of carnivorous birds also represented in the Santa Cruz Formation: *Badiostes patagonicus* (from La Cueva Locality), *Thegornis debilis* (from Puesto Estancia La Costa locality) and *T. musculosus* (from Yegua Quemada locality) (Fig. 7.1b, d, i).

Falcons are small- to medium-sized birds of prey that differ from other raptors in killing with their beaks instead of their feet (Sustaita 2008). The family has a worldwide distribution excepting the Arctic and Antarctica, and the densest forest of central Africa.

Noriega et al. (2011) corroborates the validity of *Thegornis musculosus* and its falconid affinities analyzing a very well-preserved and complete specimen of this species. Their cladistic studies confirm the phylogenetic placement within the basal clade of falcons Herpetotherinae (forest-dwelling falconids *Micrastur* and *Herpetotheres*), a peculiar group of falconids that are distributed in the lowland and mid-elevation humid forest of Central and South America (Fuchs et al. 2011).

The holotypes of *Badiostes patagonicus* and *Thegornis debilis* do not allow a more precise assignation below the family level. Interestingly, on the basis that both species of *Thegornis* have the same procedence and age and that difference in size between males and females are frequent in diurnal raptors, Noriega et al. (2011) suggested tentatively that *Thegornis debilis* could be a male of *Thegornis musculosus*.

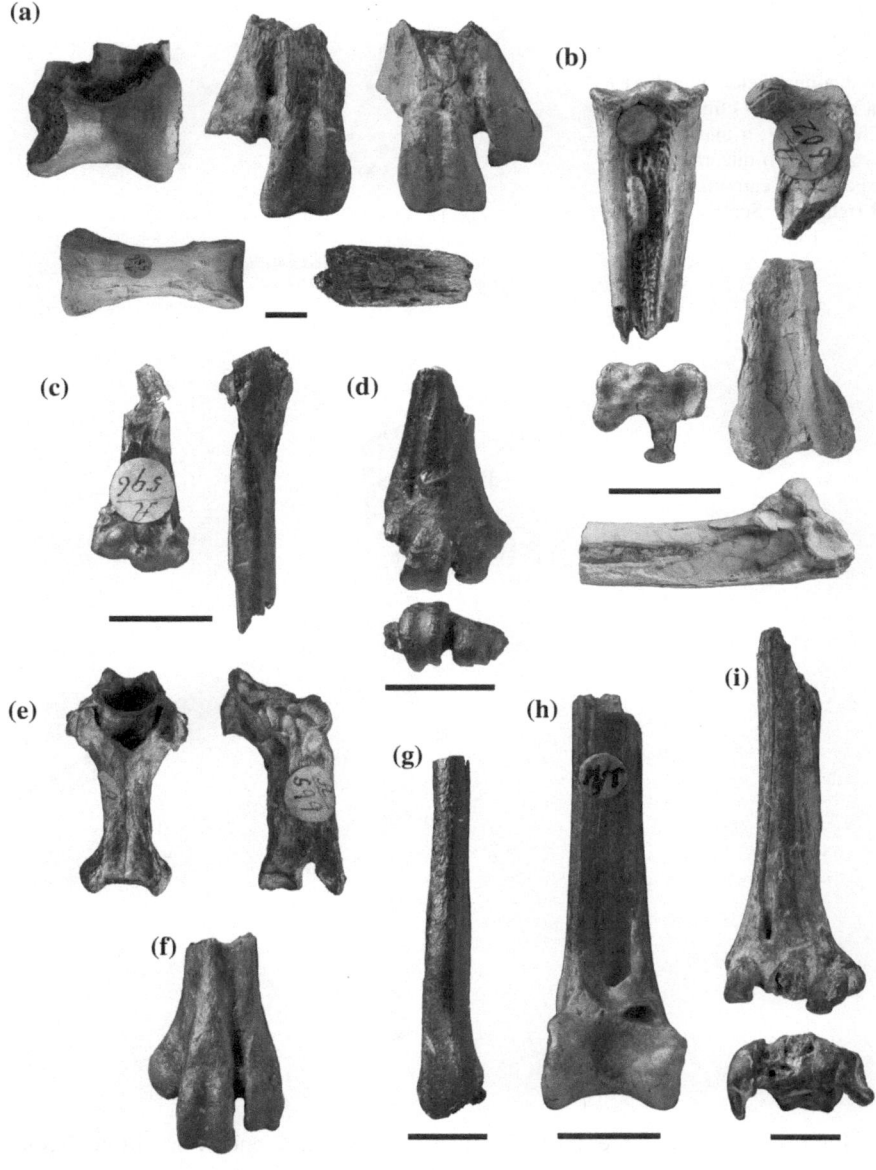

Fig. 7.1 Santacrucian fossil birds **a** *Opistodactylus patagonicus*, tibiotarsus, tarsometatarsi, phalanxes, and portion of beak (BMNH-A586-587), **b** *Badiostes patagonicus* distal portion of right tarsometatarsus (BMNH A602), **c** *Eutelornis patagonicus*, fragmentary right humerus and right tibiotarsus (BMNH A596), **d** *Thegornis debilis*, distal fragment of right tarsometatarsus (BMNH A601), **e** *Liptornis externus*, cervical vertebra (BMNH A599), **f** *Anisolornis excavatus*, distal portion of left tarsometatarsus (BMNH A594), **g** *Eoneornis australis*, fragmentary radius (BMNH A595), **h** *Protibis cnemialis*, distal portion of right tibiotarsus (BMNH A598), **i** *Thegornis musculosus* distal portion of right tarsometatarsus (BMNH A600). Scale bar = 1 cm

Fig. 7.2 *Thegornis musculosus* from the Early Miocene of the Sarmiento formation (Trelew member) in the southern cliff of the Chubut river: **a** mandibular symphysis, **b** quadrate, **c** portion of carpometacarpus, **d** right foot. Scale bar = 1 cm

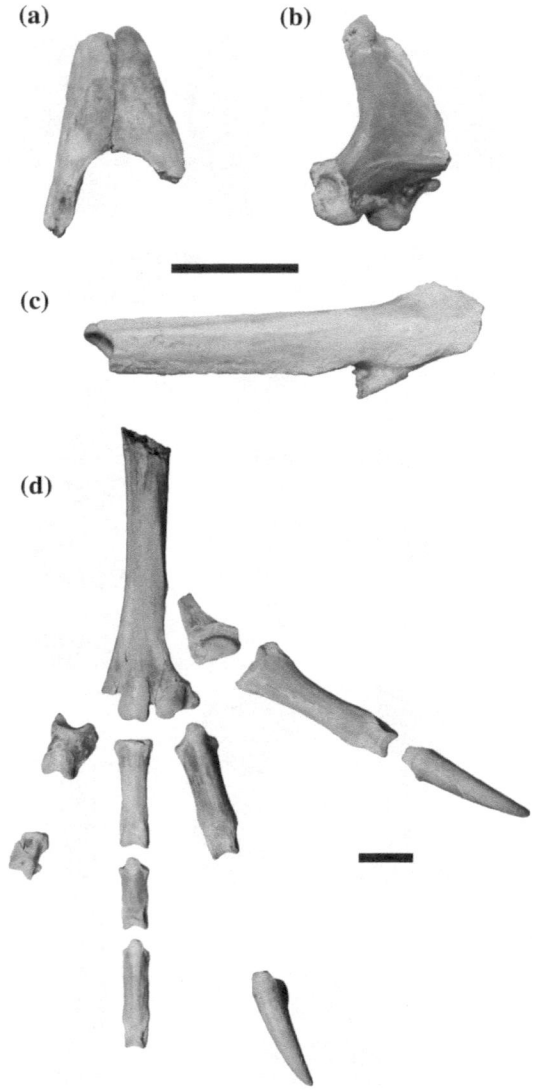

Interestingly, from the Early Miocene of the Sarmiento Formation (Trelew Member) in the southern cliff of the Chubut river near Gaiman, it was recovered a tarsometatarsus and associated pedal phalanges with other fragmentary remains (Fig. 7.2). In an initial work, Tambussi et al. (2003) proposed its similarities with the Falconidae Herpetotherinae more than any other group of falcons, although some differences with *Micrastur* could be established, e.g. the cross-section of the shaft much more anteroposteriorly compressed, the more elongated trochlea III, and a larger intertroclear internal groove. Also, the authors noted very pronounced differences between Polyborinae and Falconinae (the other two clades of the family, Fuchs et al. 2012), both in general morphology and in the arrangement of the trochleae. Now, the availability of the amazing specimen of *Thegornis musculosus* described by Noriega et al. (2011) allows us to assign the material of Gaiman to *Thegornis musculosus*. In this regard, the Gaiman specimen extends back the biochron of the taxon (Early Miocene) and the geographic distribution 900 km north of the holotype type locality.

The fossil record of darters or snakebirds (Pelecaniformes Anhingidae) dates from the Early Miocene (Fig. 2.1) and is rather abundant in South America since its first records (Cenizo and Agnolín 2010 and the literature cited therein). Anhingids are large birds of about 80–100 cm in length, with long thin necks that impale fishes with their thin pointed beaks. Currently, they have a pantropical distribution (tropical America, Africa, Asia, and Australia) but in the past they had a wide temporal and geographical distribution through North America, Europe, Africa, and Australia (Olson 1985). They frequent freshwater environments but occasionally inhabit marine coast and marshes. It is accepted that *Liptornis hesternus*, of which only a cervical vertebra is known from La Cueva fossil locality at Santa Cruz Province in Argentina (Ameghino 1895), is a valid darter species (Degrange et al. 2012). We follow here the criterion of (Degrange et al. 2012) that confirms the belonging of *L. hesternus* to the Anhingidae from a re-examination of some fossil, the examination of a cast housed in the Field Museum (Chicago) by one of the authors (FJD), and photos of the holotype (Fig. 7.1e).

Three giant-sized genera with at least six species, and a diminutive species belonging to *Anhinga* were described in the last decade (Alvarenga 1995; Alvarenga and Guilherme 2003; Areta et al. 2007; Campbell 1996; Cenizo and Agnolín 2010; Noriega 1992; Rasmussen and Kay 1992; Rinderknetch and Noriega 2002). Giant darters such as *Macranhinga* were also found in deposits of the upper Bandurrias River, Santa Cruz Province in Argentina that belongs to Santa Cruz Formation (Cenizo and Agnolín 2010). Anhingids had an important radiation since the Late Miocene at northern latitudes in Argentina, Uruguay, Brazil, Colombia, Chile, and Perú including highly specialized sympatric forms of different sizes and locomotor adaptations. Giant darters disappeared from South America in the Early Pliocene (Cenizo and Agnolín 2010; Tambussi 2011) probably due to a combination of deteriorating climatic conditions, regression of epicontinental seas, and subsequent disappearance of several freshwater environments (Cenizo and Agnolín 2010). Competition with phalacrocoracid cormorants (first fossil record during the Late Miocene marine assemblages of Perú and Chile) may also have caused the disappearance of giant darters. It is also important to note that anhingids from the Santa Cruz Formation represent the southernmost records for the family.

The Anseriformes are represented by at least four species, all of them based on fragmentary material. Using the original description made by Ameghino (1895), Cenizo and Agnolín (2010) relate the Anseriformes *Eoneornis australis* with the screamers (Anhimidae). While their arguments seem reasonable, the material is too fragmentary and little eloquent (a distal portion of a radius) to validate this assignation (Fig. 7.1g).

In this work it is considered to be an Anseriformes of uncertain affinities as suggested by Tambussi and Noriega (1996). A determination more accurate below the ordinal level is held down by the findings of new materials. This is the same situation that happens with *Eutelornis patagonicus*, the name that Ameghino gave to this species based on a distal fragment of a humerus and a proximal portion of a tibiotarsus (Fig. 7.1c). According to Cenizo and Agnolín (2010), *Eutelornis* shares plesiomorphic features with the Anseranatidae and corresponds to a basal Anseriformes. In this work, we prefer to preserve the doubtful familiar status. In the same publication, Cenizo and Agnolín (2010) described the new taxon *Ankonetta larriestrai* which is an anatid of middle size with superficial resemblance with the whistling ducks *Dendrocygna* sp. Whistling ducks are freshwater primitive anseriforms with worldwide distribution through the tropics and subtropics.

The systematic position of *Brontornis burmeisteri* has also been discussed. From its relationship with the Phorusrhacidae (Brodkorb 1967; Mourer-Chauviré 1981; Alvarenga and Höfling 2003; Alvarenga et al. 2011) and, in particular, with *Paraphysornis brasiliensis*, *Brontornis* has been related and relocated within the Anseriformes (Moreno and Mercerat 1891; Agnolín 2007; Tambussi 2011; Degrange 2012; Degrange et al. 2012). *Brontornis* is known by several remains, including hindlimbs, phalanx, vertebra, a very fragmentary quadrate supposedly associated and mainly by mandibular fragments (Fig. 7.3). Moreno and Mercerat (1891) pointed out some similarities between the hindlimb bones of *Brontornis* with those of the swan *Cygnus* (Anseriformes Anatidae) and even Dolgopol de Saez (1927) created an order apart for the genera *Brontornis* and *Rostrornis* (junior synonym of the first), based on the trochlear spread of the tarsometatarsus and the shape of the ungueal phalanxes (Fig. 7.3d, g). Agnolín (2007) proposed that this species is related with the Galloanserae and particularly with the Anseriformes. However, this hypothesis is sustained on the basis of the study of the fragmentary and isolated quadrate bone, which assignation is very questionable (Degrange 2012). In summary, *Brontornis* is not a terror bird but an Anseriformes fundamentally on the morphology of the hindlimbs, as was noted and resalted by Moreno and Mercerat (1891) and Tambussi (1989), and also based on the cladistic analysis performed by Degrange (2012) in his thesis. *Brontornis burmeisteri* Moreno and Mercerat 1891 better represented the Anseriformes during the Santacrucian Age. Its body mass was estimated as 420 kg (Degrange 2012) and it surpassed the 2 m height (Jones 2010; Degrange et al. 2012). Birds with such masses and sizes can only be associated with open habitats. *Brontornis* has a short, wide, and tall mandibular symphysis and a short but wide tarsometatarsus which reaches between 50 and 60 % of the tibiotarsus length (Alvarenga and Höfling 2003). These features led Tonni (1977) and Tambussi (1997) to propose that this bird was a scavenger of low movements. However, Agnolín (2007), based on the mandible morphology and the relationship of this taxon with the Anseriformes, proposed that *Brontornis* was herbivorous; in contraposition with Alvarenga and Höfling (2003) who hypothesized that brontornithines may have been scavengers/kleptoparasites. Nevertheless, skull remains are fragmentary and the assumption of any trophic habit with certainty is speculative (Degrange et al. 2012).

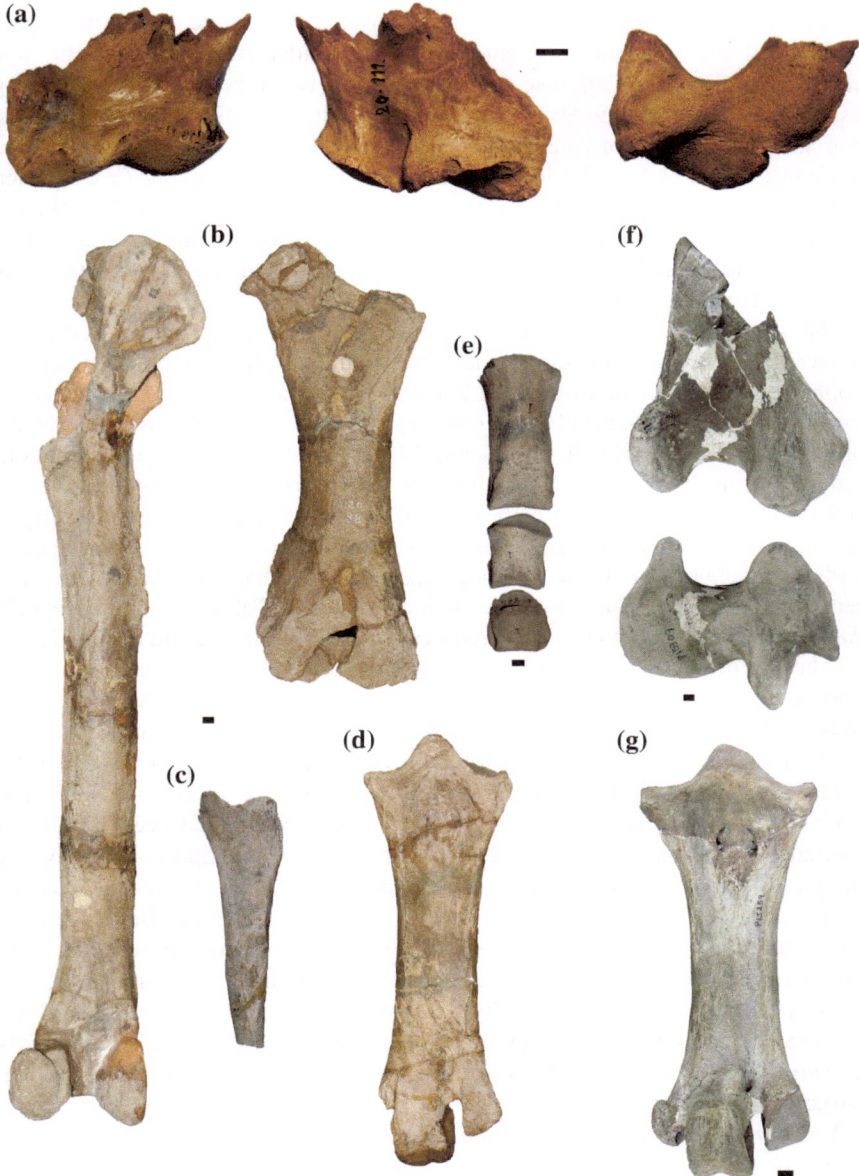

Fig. 7.3 *Brontornis burmeisteri*: **a** fragmentary quadrate (MLP 20-111) in cranial, caudal, and distal views, **b–g** hindlimb bones, **b** left femur and tibiotarsus (MLP 20-88), **c** left fibula (MLP 20-90), **d** left tarsometatarsus (MLP 20-91), **e** phalanx 1, 2, and 3 from the third toe (MLP 20-574, 20-575, 20-580), **f** distal portion of left femur (FMP15309) in cranial and distal views, **g** left tarsometatasus (FM-P15259). Scale bar = 1 cm

The Santacrucian Cariamiformes are represented by two families: Cariamidae and Phorusrhacidae. The South American seriemas constitute the single extant members of the small and ancient family Cariamidae, which is also the sole surviving family of the Cariamae. They are the only two very closely related living flesh eating species around 80 cm high that run rather than fly, and roost on trees. From the middle levels of Estancia La Costa Member of the Santa Cruz Formation (Noriega et al. 2009) in the locality of Puesto Estancia La Costa (= Corriguen Aike), a few bone fragments (two isolated distal ends of tibiotarsi of Cariaminae), and *Cariama santacrucensis* based on a fragment of neurocranium (Noriega et al. 2009) were recovered. *Cariama* is the oldest South American genus with living representatives (Tambussi 2011).

The Terror Birds (Phorusrhacidae) are the best bird group represented in the Santacrucian, both in number of species and specimens (see "The predominance of zoophagous birds"). One of the best-known taxa of the 81 species nominated by Ameghino is the phorusrhacid *Phorusrhacos longissimus* that was nominated on the basis of a jaw and originally considered for Ameghino himself as an edentulous mammal (Ameghino 1887). This bird was a terrestrial, non-flying, carnivorous predator of huge size, with a body mass of approximatelly 120 kg according to Degrange (2012). After *Phorusrhacos*, Ameghino nominated another 21 species of Phorusrhacidae (Ameghino's "Phororhacosidae" + "Pelecyornidae"), most of them based on fragmentary and little eloquent material (Table 5.1). Of these, only *Phorusrhacos longissimus* is considered a valid name (Alvarenga and Höfling 2003).

Other Phorusrhacidae found in the Santa Cruz Formation belong to the Psilopterinae and Patagornithinae subfamilies in the sense of Alvarenga and Höfling (2003). Of all the places in South America where there are remains of Psilopterinae, only in Santa Cruz Formation the coexistence of two species of this clade is recorded: *Psilopterus lemoinei* (Moreno and Mercerat 1891) and *P. bachmanni* (Moreno and Mercerat 1891). Both species had been beautifully illustrated by Sinclair and Farr (1932), although the descriptions are lax and ambiguous. Based on new findings, Degrange and Tambussi (2011) recently described in-depth and made a new diagnosis of *P. lemoinei* and Degrange et al. (2011) did the same for *P. bachmanni*, but restricted to the forelimb. The Psilopterinae occupied the role of active predators of small to medium size.

Patagornithinae are represented by *Patagornis marshi*, a terror bird extensively described by Andrews (1899), represented by numerous and abundant materials. *Patagornis* would have been a medium-sized obligate terrestrial bird (true terrestrial bird) (Degrange 2012).

Anisolornis excavatus is one of the most controversial species assignment. Its systematic position has been changing since its original description by Ameghino in 1891a. Based on a distal fragment of left tarsometatarsus (Fig. 7.1f), Ameghino was originally considered as a Psilopterinae (= Pelecyornidae) and then changed to be related to the Cracidae by Ameghino himself in 1895. Later authors adopted different criteria: Brodkorb (1964) ubicated between the Galliformes and even in the Tinamiformes; Cracraft (1973) and Olson (1985) supposed some relationship with the limpkins (Gruiformes Aramidae) and thrumpeters (Gruiformes Psophiidae).

Our study allows us to confirm its assignment to the "Gruiformes" in the classic sense, but due to the fragmentary nature of this remains, more accurate location is difficult. This does not particularly help to clarify the systematic picture about the Miocene birds. As is known, a number of wading and terrestrial bird families that did not seem to belong to any other order were classified together as Gruiformes. In this manner, Gruiformes (cranes, rails, crakes, limpkin, etc.) contained a considerable number of living and extinct bird families, with a widespread geographical diversity.

Protibis cnemialis was nominated by Ameghino in 1891 on the basis of a distal end of tibiotarsus (Fig. 7.1h) that, according to Brodkorb (1963), could correspond to a Treskiornithidae (Ciconiiformes), a criterion used in later works (Tonni 1980; Tambussi and Noriega 1996). More recent authors (e.g., Degrange et al. 2012) consider it a plataleid, the group that includes modern Spoonbills. Plataleids are large, long-legged wading birds grouped in the family Threskiornithidae, which also includes the Ibises.

In summary, the Early-middle Miocene avifauna of the Santa Cruz Formation include unquestionably phorusrhacids, seriemas, rheas, and falconiforms. Additionally, fragmentary specimens referred with doubts to pelecaniforms, anseriforms, gruiforms, and ciconiiforms are also present. Birds such us spoonbills, darters, and waterfowls, allow us to infer the presence of temporarily flooded savannas or permanent water bodies in forested areas. Birds such as rheas, tinamous, or seriemas indicate that scenarios with alternating areas of wooded or shrubby with herbaceous vegetation areas, are also possible.

7.2 Birds from the Miocene Pinturas Formation

The continental sequence of the Late Early-middle Miocene Pinturas Formation outcrops at the eastern border of the Deseado Massif, west central Patagonia. Vertebrate fossils were first collected in the Pinturas River valley by Carlos Ameghino in 1891. Additional localities have yielded more vertebrate fossils (Estancia Ana Maria, Arroyo Feo, Arroyo La Caldera, Arroyo Telken, valley of the Rio Ecker, Cañadón Caracoles, Cañadón Seco, Cañadón Olvidado, and Cerro Chato; Bown and Larriestra 1990). The age in relation to the Santacrucian is still controversial, and there is no agreement in reference to the prevailing environment. Some faunal components and palynological data suggest the presence of humid forests, whereas sedimentologic, paleopedologic, and ichnologic evidence indicate environments dominated by herbaceous vegetation (Kramarz and Bellosi 2005). A variety of birds have been reported from these deposits (Chiappe 1991; Noriega and Chiappe 1993; Bertelli and Chiappe 2005). The remains are usually isolated and eroded fragments belonging to Tinamidae (Tinamiformes), Falconidae Polyborinae (Falconiformes), Strigidae (Strigiformes), Anatidae (Anseriformes), Cariamidae (Cariamiformes), and Tyranni (Passeriformes). It is worth emphasizing that remains of Tinamidae, Strigidae, and Tyranni are the oldest records of these taxa for South America (Noriega and Chiappe 1993).

7.3 Late Miocene Birds from Cerro Azul Formation

Cerro Azul Formation (Late Miocene) is a continental sequence that is found in La Pampa and southwest Buenos Aires Provinces (Argentina). Deposits have a loess-like appearance (Goin et al. 2000), containing a mixture of reddish-brown terrestrial fine sand, minor silt, and sparse lenses of clay (Linares et al. 1980; Vezzosi 2012) with isolated and caliche-like concretions of irregular thickness in the same localities (Campbell and Tonni 1980). The well-known Salinas Grandes de Hidalgo locality, classically assigned to the "Epecuén Formation" was later included in the Cerro Azul Formation (Goin et al. 2000).

Aves from Cerro Azul Formation include few but interesting remains referred to ten taxa belonging to six families, four of which have living representatives, and two are extinct (Cenizo et al. 2011; Vezzosi 2012). This set of birds includes the oldest records of *Eudromia* and *Nothura* (Tinamidae), *Milvago* (Falconidae), *Pterocnemia* (Rheidae), and an undetermined Tyrannidae. Remains of phorusrhacids (*Procariama simplex*) and the giant teratorn *Argentavis magnificens* (Fig. 7.4) are also recorded. It is interesting to note here that *Procariama simplex* is considered the largest cursorial psilopterine predator with reduced forelimbs (Alvarenga and Höfling 2003; Vezzosi 2012). *Procariama* from Cerro Azul Formation documents a wide geographic distribution from northwestern (Andalhualá Formation, Catamarca Province) to central Argentina (Cerro Azul Formation, La Pampa Province). The species becomes extinct after this time.

Another striking species is *Argentavis magnificens*. We only mention here that this is the largest flying bird so far known and its relationship to the North American teratorns has been considered frequently. Information about this bird is expanded in the section on carnivorous birds.

The palaeornithological record from the Cerro Azul Formation is congruent with palaeoenvironmental inferences previously drawn from mammals recovered from the same unit. Possible scenarios for that moment are made of open environments, maybe xerophyllous shrubby steppes, perhaps with some forest (Cenizo et al. 2011). These records are the first indications of a typically Pampean bird fauna at the end of the Late Miocene in central-southern Argentina (Cenizo et al. 2011).

7.4 Late Miocene Birds from Puerto Madryn Formation

Puerto Madryn Formation is a sequence constituted of sediments which held a wide fauna of marine invertebrates (e.g., equinoderms, brachiopods, and bivalves). The vertebrate remains of this formation, although scarce, are very well preserved and even complete skeletons have been recovered in some cases (e.g., Acosta Hospitaleche et al. 2007a). Based on the abundant record of palynomorphs from disparate origins (terrestrial, marine, and aquatic) it can be inferred that the deposition took place on the inner continental shelf (Dozo et al. 2010).

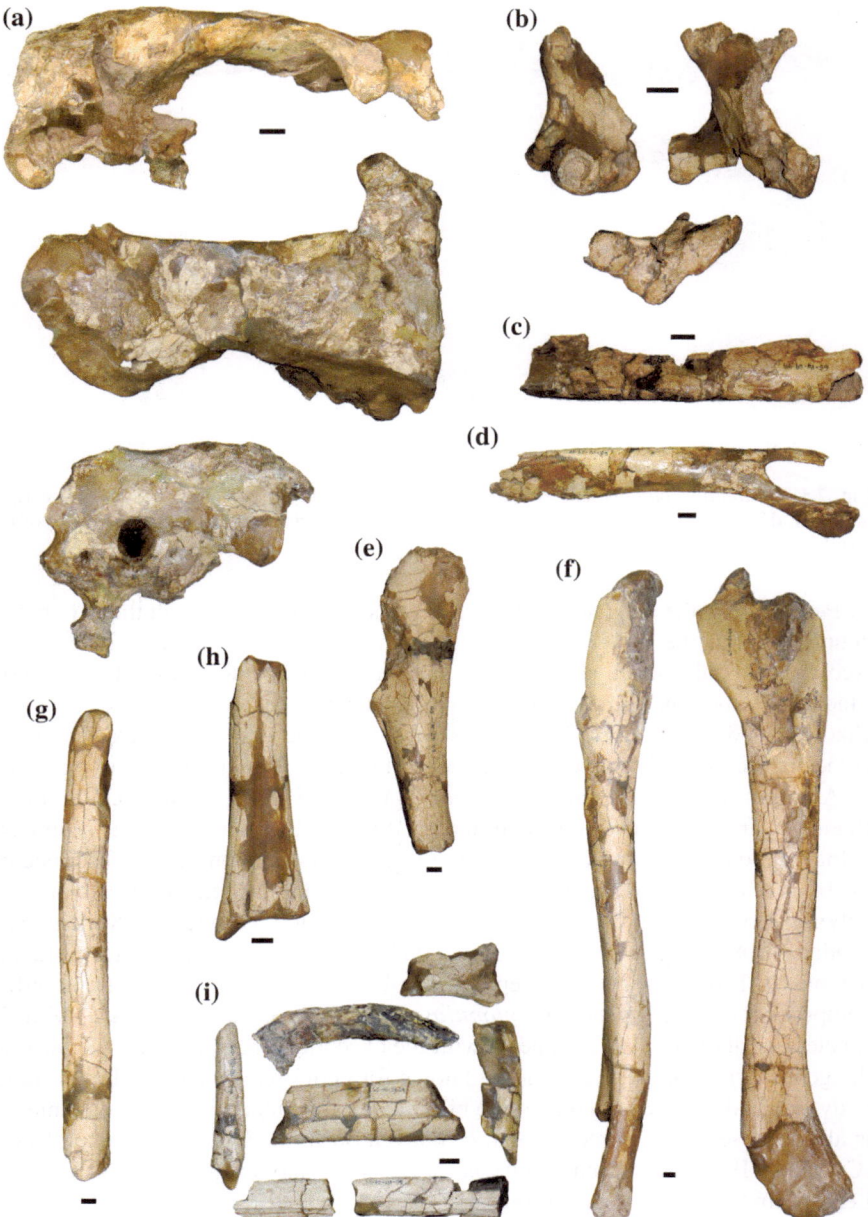

Fig. 7.4 *Argentavis magnifiscens* MLP 65-VII-29-49: **a** fragmentary skull, **b** right quadrate, **c–d** facial skull fragments, **e**, distal portion of right coracoid, **f** left humerus in lateral and cranial view, **g** shaft of right tibiotarsus, **h** shaft of right tarsometatarsus, **i** bone fragments. Scale bar = 1 cm

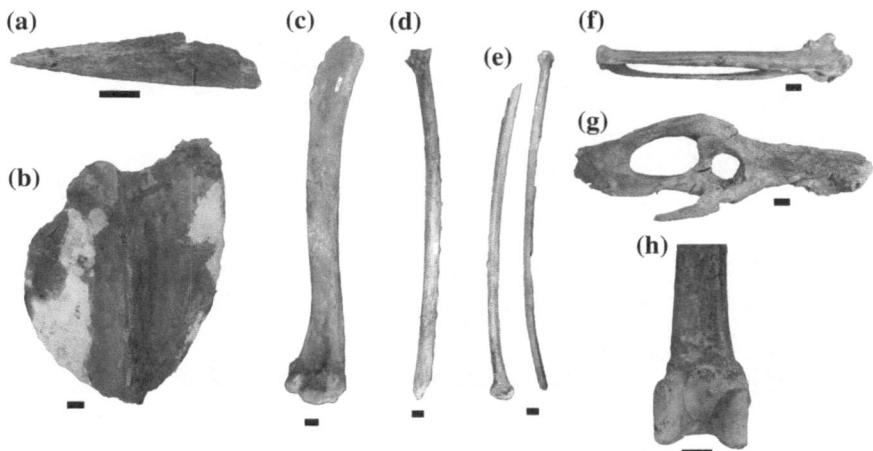

Fig. 7.5 *Leptoptilos patagonicus* MPEF-1363: **a** fragmentary symphysis, **b** sternum, **c** right humerus, **d** left ulna, **e** radii, **f** left carpometacarpous, (**g**) pelvis in lateral view, (**h**) detail of the distal portion of the left tibiotarsus. Scale bar = 1 cm

Penguins are the most abundant birds in this formation, although they are not as abundant as in the older Gaiman Formation (Acosta Hospitaleche 2003, 2004; Acosta Hospitaleche et al. 2007a; Cione et al. 2011). Recently, from Puerto Madryn Formation, some continental birds have been recovered (Noriega and Cladera 2008; Dozo et al. 2010). This constitutes the first continental vertebrate association coming from the Late Miocene of Chubut Province (Dozo et al. 2010).

A large stork *Leptoptilos patagonicus* (Fig. 7.5) was exhumed at Punta Buenos Aires (northwestern extreme of Península Valdés) from sediments belonging to the lower levels of the Puerto Madryn Formation. It was described on the basis of a partial skeleton (few fragments of the skull and mandible, tibiotarsus, pelvis, sternum, and cervical vertebrae) of a single individual (Noriega and Cladera 2008). As all Leptoptilini, *Leptoptilos patagonicus* is a large stork and it is the oldest Tertiary record of Leptoptilini for South America. The Leptoptilini comprises three living genera (*Leptoptilos, Ephippiorhynchus*, and *Jabiru*) with six species distributed on all continents with the exception of Antarctica. The birds of the genus *Leptoptilos* are commonly known for being scavengers, with their large body size and their long and massive bills. This Ciconiidae tribe in South America is also represented by a record of *Jabiru mycteria* in the Late Pleistocene of Perú (Campbell 1979). At present *Jabiru mycteria* is the only extant resident species of Leptoptilini in the Neotropical region.

Abundant remains of Anseriformes Dendrocygninae are also recorded at Puerto Madryn Formation (Acosta Hospitaleche et al. 2007b; Dozo et al. 2010). This record corresponds to the southernmost record of Dendrocyninae (Dozo et al. 2010; Tambussi 2011). The whistling ducks or tree ducks are either considered a separate

Fig. 7.6 Skull portion of the Accipitridae MPEF-PV2523: **a** dorsal, **b** ventral, **c** lateral, and **d** caudal views. Scale bar = 1 cm

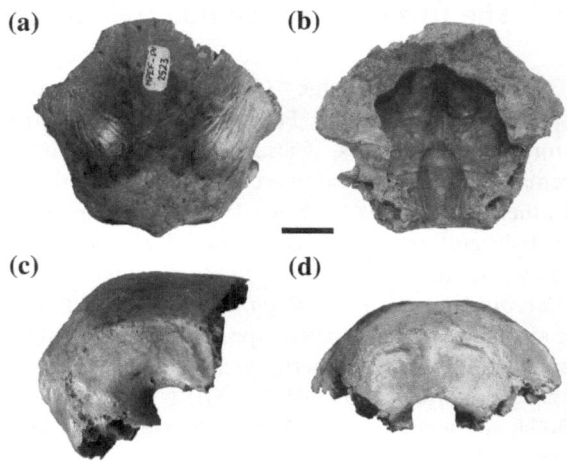

family "Dendrocygnidae" or a tribe "Dendrocygnini" in the goose subfamily Anserinae. Modern whistling ducks are herbivorous, flying, and gregarious birds with long legs and necks. They are associated with lentic environments with dense surface vegetation.

Simultaneously living at the same space are two unquestionably carnivore groups of birds: accipitrids and phorusrhacids. Both were recovered from La Pastosa and Rincón Chico localities (Dozo et al. 2010). The fossil Accipitridae belongs to a large eagle and constitutes the first skull remains available for this family in South America (Picasso et al. 2009) and corresponds to a similar animal to the extant *Geranoaetus melanoleucus* (Fig. 7.6). In Argentina, fossil Accipitridae have been known since the Eocene, represented by only a few postcranial fragments (Tonni 1980; Tambussi and Noriega 1996). Ancient fossil accipitrid records are known from the Eocene of America and Europe (Olson 1985; Mayr 2005, 2009). In America, the most abundant record comes from North America (e.g., Cracraft 1969), but it is very impoverished in South America.

Phorusrhacid Psilopterinae is represented by two unassociated remains, a cervical vertebrae and one ungual phalanx. As mentioned previously, Psilopterinae includes five species distributed from the Eocene to the Pliocene of Argentina and Brazil. They are the most graceful and smallest representatives within Phorusrhacidae and are classically associated with predatory habits (Degrange and Tambussi 2011; Degrange 2012).

The record of a Dendrocygninae is consistent with the presence of freshwater environments, while Accipitridae and Psilopterinae remains indicate the presence of open shrub.

7.5 The Ituzaingó Formation (late Miocene-early Pliocene)

The bird record from the "Conglomerado osífero" comprises more than a dozen species of seven orders: Pelecaniformes, Charadriiformes, Anseriformes, Ciconiiformes, Rheiformes, Cariamiformes, and Gruiformes. The abundant avian remains from this locality were studied by Noriega during his Ph.D. research and further publications (Noriega 1994, 1995, 2001).

Pelecaniformes are represented by several anhingids including the largest known darter *Macranhinga paranensis* with an estimated body mass of 5 kg, *Macranhinga ranzii, Anhinga minuta*, and *cf. Giganhinga* and another darter with a size similar to the recent species *Anhinga anhinga*. The former showed mechanisms of aquatic and aerial locomotion, whereas the latter have been a flightless species (Noriega 1994, 1995, 2001; Noriega and Agnolín 2008; Noriega and Piña 2004; Areta et al. 2007). Anhingids live mainly in tropical freshwater habitats and are specialized diving piscivores with elongated necks and extended pointed bills used for spearing fish (Johnsgard 1993). Fossil anhingids are frequent in South American Cenozoic deposits (Noriega 1992; Tambussi and Noriega 1996; Noriega and Alvarenga 2000; Rinderknecht and Noriega 2002; Alvarenga and Guilherme 2003; Areta et al. 2007) (Fig. 1.2).

Charadriiform records include an indeterminate species of flamingo Phoenicopteridae and another indeterminate species of the genus *Megapaloelodus* from the extinct family Palaeolidae (Noriega 1995). Recent flamingos live in swamp and floodplain environments; they are typically wading birds although some primitive fossil forms have been able to dive (Feduccia 1999).

The ciconiiform Mycteriini storks are only known in the Tertiary of South America in the "Conglomerado osífero" (Noriega 1994, 1995). Additionally, an indeterminate species of Ciconiini (tarsometatarsus and humerus of *cf. Ciconia*) is also known.

Rallidae (coots and allies) are poorly represented in this formation.

The only fossil South American representative of the Gruidae ("Gruiformes") and Dendrocheninae (Anseriformes, Anatidae) occurs in the "conglomerado osífero". Cranes are represented by two fragments of tarsometatarsi similar to those of *Grus*. Cranes have a good record from the Eocene and Oligocene of Europe, North America (Feduccia 1999), but definitely not in South America. Dendrocheninae is more related to the Dendrocygninae (whistling ducks) and Anserinae (geese and swans) than to Anatinae (ducks) (Noriega 1995). Additionally, real ducks (Anatini) also occur in Ituzaingó Formation.

Cursorial birds are represented by Rheidae and Phorusrhacidae (Noriega 1995; Tambussi and Noriega 1996).

Fossils rheid remains are common (several tibiotarsi, tarsometatarsi, and one femur and humerus) and show marked morphological affinities with *Pterocnemia* although the materials could not be assigned to any living or extinct known species until recently (Noriega and Agnolín 2008; Agnolín and Noriega 2012). A new species of the family, *Pterocnemia mesopotamica*, was recently named on the basis of a fragmentary distal portion of a right tarsometatarsus that could also be represented in the Aisol Formation (Middle to Late Miocene) of Mendoza Province, Argentina (Agnolín and Noriega 2012).

Phorusrhacids are represented by fragments of at least three species but are of doubtful assignation (Noriega and Agnolín 2008): *Devincenzia pozzi*, *Andalgalornis steulleti*, and a Phorusrhacidae indet. We agree with Noriega (2000) when he says that definitively Phorusrhacid taxa from the "Conglomerado osífero" must be revised.

Integration of data for the Ituzaingó Formation shows that the environment would have included flooded and swampy areas near some wooded areas developed around rivers and open areas of savannas and grasslands away from them (Herbst 2000; Noriega 1995; Noriega and Agnolín 2008). Both the terrestrial and freshwater fauna of the "conglomerado osífero" indicate a warmer climate than today. Freshwater vertebrates suggest important connections between the southern and northern South America basins. Records of dendrochenin anatids and palelodin flamingos show a significative biogeographic connection between South American bird faunas with those of North America and Europe (Martin 1983; Rasmussen and Kay 1992).

7.6 The Mio-Pliocene Andalhualá Formation

Andalhualá Formation (Late Miocene–Early Pliocene) of the Santa María group (northwestern Agentina) is constituted fundamentally by an assembly of upward-coarsening sandstones, with abundant conglomerates and some pelitic and tophus beds (Anzótegui et al. 2007; Bossi and Muruaga 2009). As it was previously mentioned, this is the thickest formation of the group (Marshall and Patterson 1981). It has contributed a great amount of fossil plants (Anzótegui et al. 2007) and it is the richest in fossil vertebrate remains, including fossils of birds, reptiles, and especially mammals. Herrera and Ortiz (2005) stated that the temporary boundaries for Andalhualá Formation are 7 and 3.54 Ma.

With regard to the bird remains, until now only carnivorous birds have been found (Rovereto 1914; Patterson and Kraglievich 1960; Campbell 1995; Agnolín 2006a) and one species of Palaelodidae (Nasif 1988; Agnolín 2009a) in this formation.

The Paleolodidae (Phoenicopteriformes) remains correspond to a single species; *Megapaloelodus peiranoi* is considered as a basal species within the genus (Agnolín 2009a). It is represented by postcranial material only.

The carnivorous *Argentavis magnifiscens* (Teratornithidae) is represented in this formation only by an ungual phalanx (Campbell 1995).

Agnolín (2006a) reports the presence of the accipitrid *Geranoaetus* sp. based on a distal fragment of tarsometatarsus proceeding from this formation. This record constitutes the oldest record for the genus in South America. Specifically, the wide distributed species *G. melanoleucus* is the first reported at the Miramar Formation (late Pliocene–Early Pleistocene) (Agnolín 2006a).

Phorusrhacids are quite diverse in the Andalhualá Formation. They are represented by three species: *Andalgalornis steulleti*, *Mesembriornis incertus,*, and *Procariama simplex*. The first two species are medium-sized terror birds, meanwhile *Procariama* it is a little-sized one (Degrange 2012). Without any direct evidence, pellets found in this formation were attributed to *Procariama* (Nasif et al. 2009).

The three species of terror birds are well represented in quantity of material, especially *Procariama* (Fig. 7.7a) which is also known by a nearly complete skeleton. However, except for *Andalgalornis* (Formación Andalhualá, Fig. 7.7c), the exact stratigraphic provenance of the other two species is not well specified. For example, *Mesembriornis incertus* (Fig. 7.7b) comes from the Andalhualá Formation or the Corral Quemado Formation (Patterson and Kraglievich 1960) and in the case of *Procariama* some of the remains deposited in the Field Museum (Chicago, USA) may come from the Formation or from the overlying Corral Quemado Formation according to Marshall and Patterson (1981); meanwhile the remains deposited in the Museo Argentino de Ciencias Naturales "Bernardino Rivadavia" (Buenos Aires, Argentina) described by Rovereto (1914) may proceed from the "piso araucanense" of the indeterminate level, although they could come from the Andalhualá Formation according to Patterson and Kraglievich (1960). Recently, Vezzosi (2012) described additional material of *Procariama simplex* unequivocally proceeding from the Andalhualá Formation. Thus, this flightless psilopterine has had an extended geographic distribution from the northwest (Catamarca Province) to central Argentina (Cerro Azul Formation, La Pampa Province).

7.7 The Mio-Pliocene Pisco Formation at Perú

Due to its abundant marine vertebrate fauna, the Pisco Formation is a famous locality exposed along the Pacific coast at Perú. It is a marine sedimentary sequence formed since the Middle Miocene to the Late Pliocene (14.0–2.0 Ma). Deposits consist of tuffaceous sandy siltstones, medium and coarse-grained sandstones, shelly sandstones, conglomerates, and coquines. The environment was interpreted as littoral close to shore (de Muizon and DeVries 1985). Birds from the Pisco Formation mainly include marine birds such as Spheniscidae, Sulidae, Pelagornithidae, Laridae, Scolopacidae, Procellariidae, Diomedeidae, Pelecanidae, and continental ones such as Phalacrocoracidae, Vulturidae, and Ciconiidae (de Muizon 1981; de Muizon and DeVries 1985; Cheneval 1993; Stucchi 2003; Urbina and Stucchi 2005a, b).

The abundance, diversity, and preservation of fossil birds in these deposits are remarkable. Following the guidelines we adopted in this work, we will make some brief comments on the non-marine birds of this assemblage.

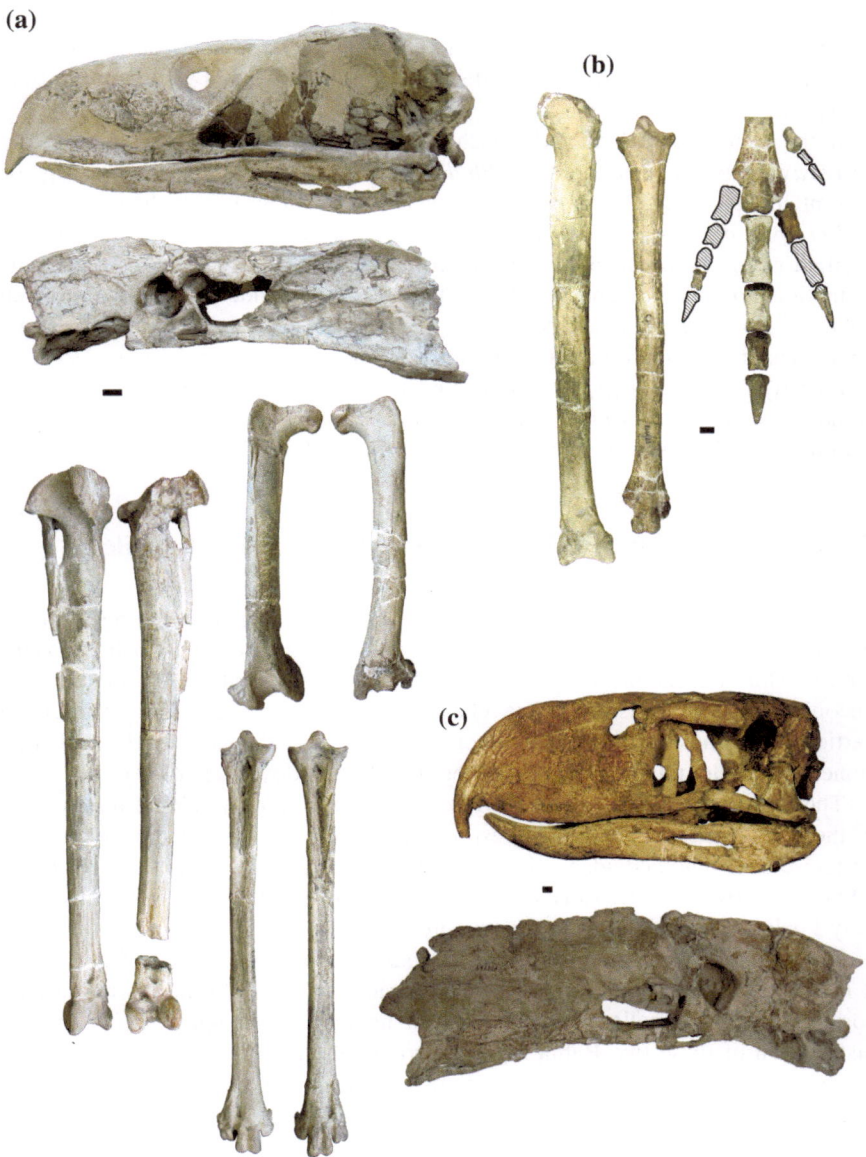

Fig. 7.7 Phorusrhacids from Catamarca province, Argentina. **a** Skull, pelvis, and hindlimbs of *Procariama simplex* FM-P14525, **b** tibiotarsus, tarsometatarsus, and right foot of *Mesembriornis incertus* FM-P14422, **c** skull and pelvis of *Andalgalornis steulleti* FM-P14357. Scale bar = 1 cm

Storks (Ciconiidae) were represented at Pisco Formation by an isolated tarso-metatarsus. Based on the environmental and trophic requirements of living storks, Urbina and Stucchi (2005a) assume that this was an occasional visitor and thus, an unexpected fossil in the area.

In the same year, Urbina and Stucchi (2005b) recognized two species of *Phalacrocorax* at Pisco Formation. One, *Phalacrocorax aff. bougainvillii* was a big cormorant and the other, *Phalacrocorax* sp. was 75 % smaller in size than the former.

Perugyps diazi is the first fossil condor (Cathartidae) described from the Pisco Formation and is the oldest condor described for South America. From the same formation, remains asssigned to a Cathartidae gen. et sp. indet are reported (Stucchi 2008). Eight fossil condors and condor-like vultures have been described for the Americas; among them four are known from South America (*Perugyps*, *Dryornis*, *Geronogyps* and *Wingegyps*) as indicated by Stucchi and Emslie (2005). These authors assume that *Perugyps* was a scavenger feeding on the carcasses of marine mammals or alternatively, it could feed on chicks of seabirds nesting on the site.

7.8 The Mio-Pliocene Bahía Inglesa Formation at Chile

The marine Bahía Inglesa Formation represents a shallow marine setting deposited within 10 km of the shore (Marquardt et al. 2000; Walsh 2002; Walsh and Naish 2002). This Formation is a clastic sedimentary sequence, fossiliferous and unconsolidated of coastal marine character, which presents strong lateral and vertical variations of facies. Lithofacies of clams, sandstones, marls, and mudstones are predominating (Marquardt et al. 2000; Godoy et al. 2003).

The birds described so far in the Pisco Formation show a marked resemblance to those reported from Bahía Inglesa in Chile. Both areas share the presence of Spheniscidae, Diomedeidae, Sulidae, Phalacrocoracidae, and Pelagornithidae (Walsh and Hume 2001; Emslie and Guerra 2003). For the purpose of this work, only Phalacrocoracidae will be commented upon.

The two cormorants reported from Perú seem to have been also in Chile. While it is not possible to argue that they are the same species, the fact is that a large cormorant and a smaller one lived simultaneously. It is highly likely that the distribution of these two Bahía Inglesa Formation cormorants extended 1600 km north, from Chile to Perú. The same seems to happen with some other marine birds (e.g., spheniscids, pelagornithids, and sulids).

7.9 Birds from the Chapadmalal Formation

Concerning the Pliocene, birds from the Chapadmalal Formation, between Mar del Plata and Miramar, southeastern Buenos Aires (Argentina) include Rheidae, Tinamidae, Phorusrhacidae (Psilopterinae and Mesembriornithinae), Vulturidae, Charadriidae, Scolopacidae, and Furnariidae (Tambussi 2011).

Rheids are terrestrial and cursorial birds and are now the greater birds of America, endemic to the Neotropical region with a definite South American origin which never left the continent. *Hinasuri nehuensis* was collected in the upper cliffs of Monte Hermoso ("Chapadmalalense" of Vignati 1925, Middle-late Pliocene, see Deschamps et al. 2011 for correlations between Monte Hermoso and Chapadmalal areas). The left femur (MLP 86-VI-21-1) on which Tambussi (1995) described the taxon, is far more robust than in the known extant and fossil species of rheas (Rheiformes) (Tambussi 1995). The Rheidae seem to have been very conservative since its earliest records, maintaining the overall pattern and proportions of the living species except in *Hinasuri*. It is worthy to mention here the presence of another robust rheid in the Plocene of Monte Hermoso locality, *Heterorhea dabbenei* (Cenizo et al. 2011; Agnolín and Noriega 2012) whose holotype is missing.

Tinamous (Tinamiformes) are mainly ground-dwelling birds whose fossil records occur primarily during the Pliocene and Pleistocene of Argentina (Picasso and Degrange 2009). They are Neotropical birds, most of which are restricted to tropical lowlands in South America. Remains of tinamous are frequent in the Chapadmalal Formation being *Nothura parvula* the most common taxon (Tambussi and Noriega 1996). Another taxon recorded is the Darwin's Nothura (*Nothura darwini*), a living species that currently inhabits the arid steppes. Fossil of the living species *Eudromia elegans* (Tinamidae) was also found at Chapadmalal Formation. The elegant crested tinamou *Eudromia elegans* is a charismatic terrestrial bird of windswept Patagonia. Nowadays it inhabits Andean steppes and mountainsides from Patagonia to Buenos Aires Province and from central to northwestern Argentina. A detailed bioclimatic analysis of its geographic range (Echarri et al. 2008) indicates low precipitation (mean of 311.45 mm) featuring its distributional areas. This is important as proxy information in paleoenvironmental reconstructions.

One species of Phorusrhacidae so far described from the Chapadmalalan is known as *Mesembriornis milneedwarsi* Moreno 1889. Like other phorusrhacids, *Mesembriornis* have ineffectual wings unsuitable to fly. It ($\sim$70 kg of body mass) was one of the taller phorusrhacids. Without any biomechanical study, Tonni and Noriega (1998) proposed that this species would have been a scavenger, competing with the fossil vultures founded in the same formation. Using a mechanical model based on tibiotarsal strength, Blanco and Jones (2005) proposed that *Mesembriornis* could have used its legs to break long bones and accessed to the marrow. However, some of the assumptions made by Blanco and Jones (2005) (e.g., femur orientation) throw some doubts on their results.

A splendid specimen of Mesembriornithinae (Degrange et al. 2011), currently being studied by us, is added to the knowledge of the Chapadmalalan birds. The material (MMP 5050) was recovered from the locality La Estafeta ($\sim$3.3 MA, Schultz et al. 1998) and allows us to know for the first time, especially the structures for Phorusrhacidae, such as ossified ring of the trachea, sclerotic rings of both orbits, ossification of the lacrimojugal ligament (os lacrimojugale communicans, a characteristic shared only with their sister group, Cariamids), furcula, and ossified medial posterior tendons associated with the tarsometatarsus. Essentially, MMP 5050 preserves the full palate corroborating the presence of vomer and palatines rostral processes. The new taxon would have reached a size smaller than that of *M. milneedwarsi*.

The Psilopterinae in the Chapadmalalan stage is represented by a Psilopterinae gen. et sp. indet. according to Tambussi (1989). The remains include a single skull with part of the neurocranium, beak, nostrils, and scleral rings preserved (Tambussi 1989; Tambussi and Noriega 1996). Unfortunatelly, the specimen was not available for this study.

Phorusrhacidae Phorusrhacinae and Physornithinae are absent in the Chapadmalal Formation. Thus, the carnivorous role during the Chapadmalalan would have been by representing necessarily terrestrial Phorusrhacids and flying condors.

Tonni and Noriega (1998) reported a fossil of the living Andean Condor from the early Chapadmalalan (4 Ma) of Río Quequén Salado (Buenos Aires) locality at Argentina. It is important to clarify that this material has not the same origin as the above specimens but the locality where it was found is geographically very close. Currently, condors (Ciconiiformes, Cathartidae) do not live in the Pampas region, but their fossil record is abundant since the Pliocene and especially during the Pleistocene (Tambussi and Noriega 1999; Tonni and Noriega 1998) in the area. They are large broad-winged soaring birds whose oldest record is from the Late Miocene/Early Pliocene (*Perugyps diazi*). In addition, an undescribed condor also from the Middle Pliocene of Argentina is known by only a proximal ulna and radius (Tonni and Noriega 1998; Tambussi and Noriega 1999). Condors are also represented by *Dryornis pampeanus* (Montehermosan, Early to middle Pliocene) based on a distal fragment of humerus and erroneously asigned to Phorusrhacidae by Moreno and Mercerat (1891) in its original description. *Dryornis*, *Vultur gryphus* and the undescribed condor mentioned above all come from the Early–middle Pliocene (Tonni and Noriega 1998; Tambussi and Noriega 1999) and represent the oldest record of condor-like Cathartidae for the Pampean Region of Argentina. Finally, from Neogene sediments at the northwest of La Rioja Province (Quebrada de la Troya, Argentina) it was recovered a distal end of humerus, quite eroded, which may correspond to *Dryornis* (Brizuela 2004). If this assignment is confirmed, then the distribution of this form would extend at least 800 km north of the Pampean Region.

It is well known that condors are soaring scavengers with high orographic affinity. For the Montehermosan/Chapadmalalan stages, arid or semi-arid conditions compatible with grasslands and forest patches have been inferred. Such conditions favor the presence of updrafts as those used by condors for soaring. As no other scavenging species are known among the fauna of the Chapadmalal Formation, it is reasonable to assume that condors filled this niche.

The waterbirds Charadriiformes *Calidris* and *Charadrius* present in the Chapadmalalan indicate the presence of freshwater environments.

Finally, the passerines are represented by the ovenbirds (Furnariidae) (Tonni and Noriega 2001). They are small- to medium-sized insectivores. Most of them are forest birds, but some are in more open habitats such as savannah or grassland.

Truly terrestrial birds such as Rheidae, Psilopterinae, and Mesembriornithinae testified the presence of open environments.

(a)

(b)

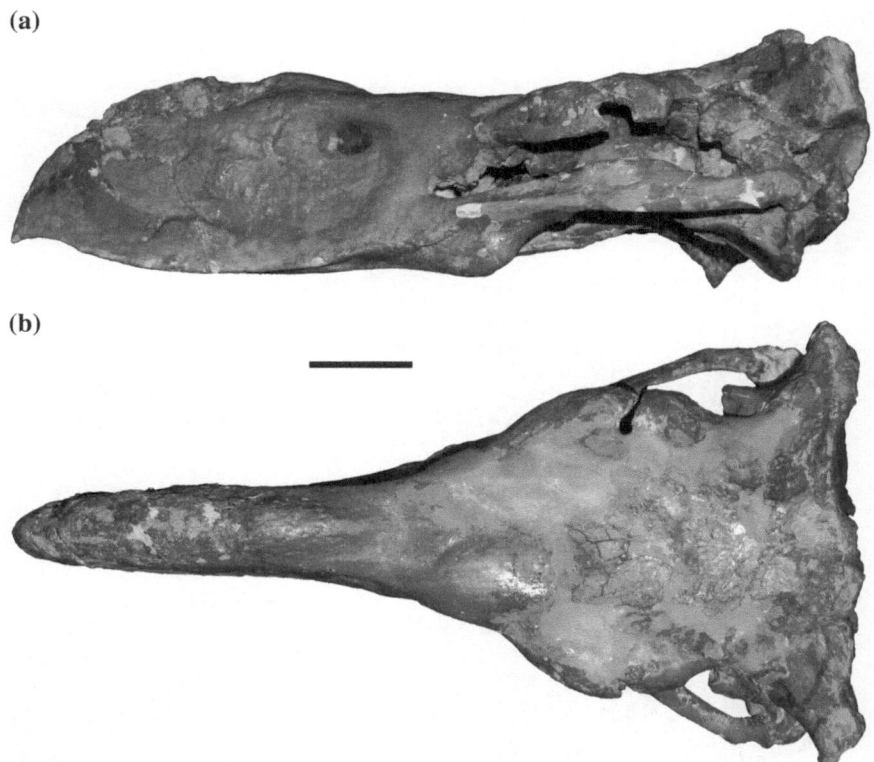

Fig. 7.8 Skull of *Devincenzia pozzi* MLP 37-III-7-83 in **a** lateral and **b** dorsal views. Scale bar = 10 cm

7.10 Significant Isolated Remains

Finally, we conclude this section with comments on some isolated records that we considered important to include here. South American Psittaciformes (parrots, cockatoos, macaws) are recognized from the Late Pliocene and all records are limited to the Pampean Region in Argentina (Tambussi 2011). All but one (*Nandayus vorohuensis* described by Tonni and Noriega (1996) are species of the genus *Cyanoliseus* and Pleistocene in age (Acosta Hospitaleche and Tambussi 2006). The holotype of *Nandayus vorohuensis* (skull and mandible MLP 94-IV-1-1) comes from "Vorohue Formation" exposed at Marquesado beach close to Miramar that is considered as Late Pliocene in age. However, Isla and Espinosa (2009) indicate that those sediments belong to the lower Pleistocene. If this were the case, the record of Psittaciformes is limited to the Pleistocene. The living representative of the genus *Nandayus* is the monotypic species of Black-hooded Parakeet, *Nandayus nenday*, distributed from southeast Bolivia to southwest Brazil, central Paraguay, and northern Argentina.

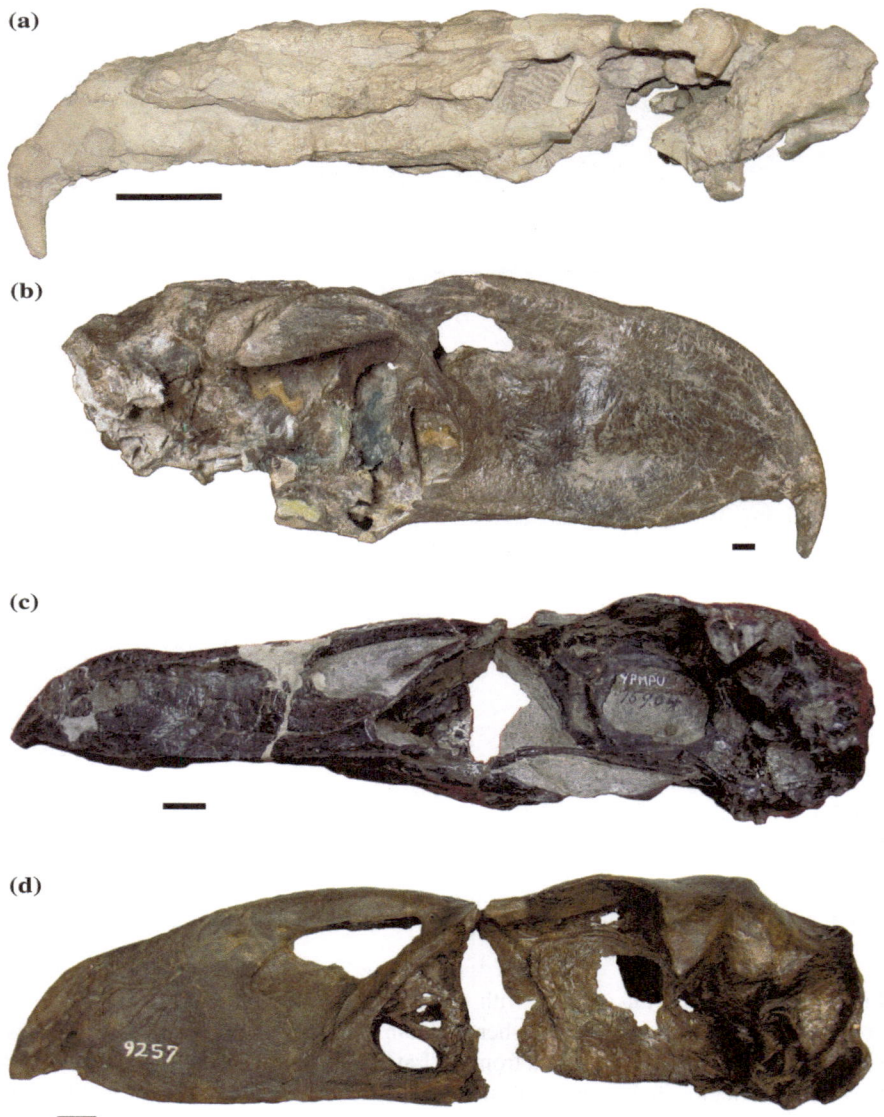

Fig. 7.9 Skull morphology of the terror birds: **a** *Kelenken guillermoi* BAR 3877-11, scale bar = 10 cm. **b** *Patagornis marshi* BMNH-A516, **c** *Psilopterus bachmanni* YPM-PU15904, **d** *Psilopterus lemoinei* AMNH9257. Scale bar = 1 cm

The large-bodied phorusracid *Devincenzia pozzi*, (originally called *Onactornis depressus*), lived in the Pampean Region during the Pliocene. A very fragmentary skull (MLP 37-III-7-83) was extracted close to Adolfo Alsina in Buenos Aires Province (Fig. 7.8). The skull fragment is currently preserved embedded in plaster, simulating the entire skull. The reconstruction housed at La Plata Museum, does not properly allow to discriminate the original pieces of those which are not. Adittional isolated materials of the species are known from Entre Ríos Province (Argentina) (Noriega and Agnolín 2008).

Another giant terror bird is *Kelenken quillermoi* (Fig. 7.9a) whose skull reaches a length of 71.6 cm and the whole animal would reach 3 m high (Chiappe and Bertelli 2006). *Kelenken* is also represented by a tarsometatarsus and a broken phalanx (Bertelli et al. 2007) and proceeds from the locality of Comallo (Río Negro Province, Argentina), where the Middle Miocene Collón Curá Formation outcrops. This terror bird is the largest known phorusrhacid.

Known only from a single incomplete skeleton that includes parts of the jaw, arm, and leg, *Psilopterus colzecus* was described by Tonni and Tambussi (1988). The holotype (MLP-76-VI-12-2) was collected at the Vivero Member of the Chasicó Formation (Late Miocene, Buenos Aires Province). It is of similar size that the Santacrucian *P. lemoinei*.

Seriemas are sister taxa of phorusrhacids, that in contrast to the latter, have a very poor record. Tibiotarsus and tarsometatarsus fragments from late Pliocene sediments of south eastern Buenos Aires Province (Argentina) allowed Tonni (1974) to recognize the extinct species *Chunga incerta*. The latter is cogenneric with the extant Black-legged Seriema (*Chunga burmeisteri*) that is found in northwest Argentina and Paraguay. The fossil record at Buenos Aires indicates a more southern and eastern distribution of the genus *Chunga* during Late Pliocene times.

References

Acosta Hospitaleche C (2003) *Paraptenodytes antarcticus* (Aves: Spheniscisformes) en la Formación Puerto Madryn (Mioceno tardío temprano), provincia de Chubut. Argentina Rev Esp Pal 18:179–183

Acosta Hospitaleche C (2004) Los pingüinos (Aves, Spheniscisformes) fósiles de Patagonia. Sistemática, biogeografía y evolución. Dissertation, Universidad Nacional de La Plata

Acosta Hospitaleche C, Tambussi CP (2006) *Cyanoliseus patagonopsis* nov. sp. (Aves, Psittaciformes) del Pleistoceno de Punta Hermengo, provincia de Buenos Aires. Ameghiniana 43:249–253

Acosta Hospitaleche C, Tambussi CP, Donato M, Cozzuol M (2007a) A new Miocene penguin from Patagonia and a phylogenetic analysis of living and fossil species. Acta Paleontol Pol 52:299–314

Acosta Hospitaleche C, Tambussi CP, Dozo MT (2007b) *Dendrocygna* (Anseriformes) en el Mioceno tardío de la Formación Puerto Madryn (Argentina): anatomía de la pelvis. Ameghiniana 44:4R

Agnolín FL (2004) La posición sistemática de algunas aves fósiles deseadenses (Oligoceno Medio) descriptas por Ameghino en 1899. Rev Mus Arg Cienc Nat 6:239–244

Agnolín FL (2006a) Notas sobre el registro de Accipitridae (Aves, Accipitriformes) fósiles argentinos. Stud Geol Salmant 42:67–80

Agnolín FL (2006b) Posición sistemática de algunas aves fororracoideas (Ralliformes; Cariamae) Argentinas. Rev Mus Arg Cienc Nat 8:27–33

Agnolín FL (2007) *Brontornis burmeisteri* Moreno y Mercerat, un Anseriformes (Aves) gigante del Mioceno Medio de Patagonia, Argentina. Rev Mus Arg Cienc Nat 9:15–25

Agnolín FL (2009a) Una nueva especie del genero *Megapaloelodus* (Aves: Phoenicopteridae: Palaelodinae) del Miocenos superior del noroeste de Argentina. Rev Mus Arg Cienc Nat 11:23–32

Agnolín FL (2009b). Sistemática y filogenia de las aves fororracoideas (Gruiformes: Cariamae). Monografías Fundación Azara

Agnolín F, Noriega JI (2012) Una nueva especie de ñandú (Aves: Rheidae) del Mioceno tardío de la Mesopotamia. Ameghiniana 49:236–246

Alvarenga HMF (1995) A large and probably flightless anhinga from the Miocene of Chile. Cou For Senck 181:149–161

Alvarenga HMF, Guilherme E (2003) The Anhingas (Aves: Anhingidae) from the upper Tertiary (Miocene–Pliocene) of southwestern Amazonia. J Vert Pal 23:614–621

Alvarenga HMF, Höfling E (2003) Systematic revision of the Phorusrhacidae (Aves: Ralliformes). Pap Av Zool 43:55–91

Alvarenga HMF, Chiappe LM, Bertelli S (2011) Phorusrhacids: the terror birds. In: Dyke G, Kaiser G (eds) Living dinosaurs: the evolutionary history of modern birds. Wiley, Chichester

Ameghino F (1887) Enumeración sistemática de las especies de mamíferos fósiles coleccionados por Carlos Ameghino en los terrenos Eocenos de la Patagonia austral y depositados en el Museo de La Plata. Bol Mus La Plata 1:1–26

Ameghino F (1889) Los mamíferos fósiles de la República Argentina. Actas Acad Nac Cienc Córd 6:1–1028

Ameghino F (1891a) Enumeración de las aves fósiles de la República Argentina. Rev Arg Hist Nat 1:441–453

Ameghino F (1891b) Mamíferos y aves fósiles argentinas: especies nuevas, adiciones y correcciones. Rev Arg Hist Nat 1:240–259

Ameghino F (1895) Sobre las aves fósiles de Patagonia. Bol Inst Geog Arg 15:501–602

Ameghino F (1899) Sinopsis geológico-paleontológica de la Argentina. Segundo Censo Rep Arg 1:112–255

Andrews C (1899) On the extinct birds of Patagonia. I. The skull and skeleton of *Phororhacos inflatus* Ameghino. Trans Zool Soc Lond 15:55–86

Anzótegui LM, Garralla SS, Herbst R (2007) Fabaceae de la Formación El Morterito, (Mioceno Superior) del valle del Cajón, provincia de Catamarca, Argentina. Ameghiniana 44:183–196

Areta JI, Noriega JI, Agnolín FL (2007) A giant darter (Pelecaniformes: Anhingidae) from the Upper Miocene of Argentina and weight calculation of fossil Anhingidae. Neu Jah Geol Pal Ab 243:343–350

Bertelli S, Chiappe LM, Tambussi CP (2007) A new phorusrhacid (Aves, Cariamae) from the middle Miocene of Patagonia, Argentina. J Vert Pal 27:409–419

Bertelli S, Chiappe L (2005) Earliest tinamous (Aves: Palaeognathae) from the Miocene of Argentina and their phylogenetic position. Contrib Sci 502:1–20

Blanco RE, Jones WW (2005) Terror birds on the run: a mechanical model to estimate its maximum running speed. Proc Roy Soc B 272:1769–1773

Bossi GE, Muruaga CM (2009) Estratigrafía e inversión tectónica del 'rift' neógeno en el Campo del Arenal, Catamarca, NO Argentina. And Geol 36:311–341

Bown TM, Larriestra CN (1990) Sedimentary paleoenvironments of fossil platyrrhine localities, Miocene Pinturas Formation, Santa Cruz Province, Argentina. J Hum Evol 19:87–119

Brizuela R (2004) Registro de Vulturidae (Aves, Ciconiiformes) en el Neógeno de La provincia de la Rioja, Argentina. Rev Mus Arg Cienc Nat 6:307–311

Brodkorb P (1964) Catalogue of fossil birds. Part II (Anseriformes through Galliformes). Bull Fla State Mus Biol Sci 8:195–335

Brodkorb P (1963) A giant flightless bird from the Pleistocene of Florida. Auk 80:111–115

Brodkorb P (1967) Catalogue of fossil birds, Part III (Ralliformes, Ichthyornithiformes, Charadriiformes). Bull Fla State Mus Biol Sci 2:99–220

Campbell KE Jr (1979) The non-passerine Pleistocene avifauna of Talara Tar Seeps, northwestern Perú. Life Sci Cont, Roy Ont Mus 118:1–203

Campbell KE Jr (1995) Additional specimens of the giant teratorn, *Argentavis magnificens*, from Argentina (Aves: Teratornithidae). C For Senck 181:199–201

Campbell KE Jr (1996) A new species of giant Anhinga (Aves: Pelecaniformes: Anhingidae) from the upper Miocene (Huayquerian) of amazonian Perú. Contrib Sci 460:1–9

Campbell KE Jr, Tonni EP (1980) A new genus of teratorn from the Huayquerian of Argentina (Aves: Teratornithidae). Contrib Sci 330:59–68

Cenizo MM, Agnolín FL (2010) The southernmost records of Anhingidae and a new basal species of Anatidae (Aves) from the lower–middle Miocene of Patagonia, Argentina. Alcheringa 34:1–22

Cenizo MM, Tambussi CP, Montalvo CI (2011) Late Miocene continental birds from the Cerro Azul formation in the Pampean region (central-southern Argentina). Alcheringa 0–22

Chandler RM (2012) A new species of Tinamou (Aves: Tinamiformes, Tinamidae) from the Early-Middle Miocene of Argentina. Palarch's J Vert Pal 9:1–8

Cheneval J (1993) L'avifaune Mio-Pliocène de la formation Pisco (Pérou) étude préliminaire. Docum Lab Géol 125:85–95

Chiappe LM (1991) Fossil birds from the Miocene Pinturas formation of southern Argentina. J Vert Pal 17:21–22A

Chiappe LM, Bertelli S (2006) Skull morphology of giant terror birds. Nature 443:929

Cione AL, Cozzuol MA, Dozo MT, Hospitaleche CA (2011) Marine vertebrate assemblages in the Southwest Atlantic during the Miocene. Biol J Linn Soc 103:423–440

Cracraft J (1969) Systematics and evolution of the Gruiformes (class Aves). 1. The eocene family Geranoididae and the early history of the Gruiformes. Am Mus Novit 2388:1–41

Cracraft J (1973) Systematics and evolution of the Gruiformes (Class Aves) 3. Phylogeny of the suborder Grues. Bull Am Mus Nat Hist 151:1–127

de Muizon C (1981) Les vertébrés fossiles de la Formation Pisco (Pérou). Première partie. Mém Inst Franç Études And 6:1–150

de Muizon C, DeVries TJ (1985) Geology and paleontology of late Cenozoic marine deposits in the Sacaco area (Peru). Geol Rund 74:547–563

Degrange FJ (2012) Morfología del cráneo y complejo apendicular posterior de aves fororracoideas: implicancias en la dieta y modo de vida. Dissertation, Universidad Nacional de La Plata

Degrange FJ, Tambussi CP (2011) Re-examination of *Psilopterus lemoinei* (Moreno and Mercerat 1891), a late early Miocene little terror bird from Patagonia (Argentina). J Vert Pal 31:1–13

Degrange FJ, Tambussi CP, Scaglia F, Dondas A, Taglioretti ML (2011) Hallazgo de un esqueleto completo y articulado de un nuevo Phorusrhacidae (Aves) en el Plioceno tardío de la Argentina. Actas IV Cong Lat Pal Vert

Degrange FJ, Noriega JI, Areta JI (2012) Diversity and paleobiology of the santacrucian birds. In: Vizcaíno SF, Kay RF, Bargo MS (eds) Early Miocene paleobiology in Patagonia: high-latitude paleocommunities of the Santa Cruz formation. Cambridge University Press, Cambridge

Deschamps CM, Vucetich MG, Verzi DH, Olivares AI (2011) Biostratigraphy and correlation of the Monte Hermoso formation (early Pliocene, Argentina): the evidence from caviomorph rodents. J South Am Earth Sci 35:1–9

Dolgopol de Saez M (1927) Las aves corredoras fósiles del Santacrucense. An Soc Cient Arg 103:145–164

Dozo T, Bouza P, Monti A, Palazzesi L, Barreda V, Massaferro G, Scasso R, Tambussi CP (2010) Late Miocene continental biota in northeastern Patagonia (Península Valdés, Chubut, Argentina). Palaeogeog Palaeoecol 297:100–109

Echarri F, Tambussi CP, Acosta Hospitaleche C (2008) Predicting the distribution of the crested-tinamous, *Eudromia* spp. (Aves, Tinamiformes). J Ornit 150:75–84

Emslie S, Guerra C (2003) A new species of penguin (Spheniscidae: *Spheniscus*) and other birds from the late Pliocene of Chile. Proc Biol Soc Wash 116:308–316

Feduccia A (1999) 1,2,3–2,3,4: Accommodating the cladogram. PNAS 96:4740–4742

Fuchs J, Chen S, Johnson JA, Mindell DP (2011) Pliocene diversification within the South American Forest falcons (Falconidae: *Micrastur*). Mol Phylogen Evol 60:398–407

Fuchs J, Johnson J, Mindell D (2012) Molecular systematics of the caracaras and allies (Falconidae: Polyborinae) inferred from mitochondrial and nuclear sequence data. Ibis 154:520–532 (article first published online: 5 March 2012)

Godoy E, Marquardt C, Blanco N (2003) Carta Caldera, Región de Atacama. Ser Geol Básica 76:1–39

Goin F, Montalvo CI, Visconti G (2000) Los marsupiales (Mammalia) del Mioceno superior de la Formación Cerro Azul (provincia de La Pampa, Argentina). Estudios Geológicos 56:101–126

Hatcher JB (1903) Narrative of the expedition. In: Scott WB (ed) Reports of the Princeton University expeditions to Patagonia, 1896–1899, vol 1: Narrative and Geography. Princeton University Press, New Jersey

Herbst R (2000) La Formación Ituzaingó (Plioceno). Estratigráfica y distribución. Corr Geol 14:181–190

Herrera CM, Ortiz PE (2005) Nuevos registros de mamíferos para la Formación Andalhualá (Mioceno tardío-Plioceno), provincia de Tucumán. Ameghiniana 42:33R

Isla FI, Espinosa M (2009) Stratigraphy, tectonic and paleogeography of the Loberia coastline, Southeastern Buenos Aires. Rev Asoc Geol Arg 64:557–568

Johnsgard PA (1993) Cormorants, darters, and pelicans of the world. Smithsonian Institution Press, Washington

Jones WW (2010) Nuevos aportes sobre la paleobiología de los fororrácidos (Aves: Phorusrhacidae) basados en el análisis de estructuras biológicas. Dissertation, Universidad de Ciencias

Kramarz AG, Bellosi ES (2005) Hystricognath rodents from the Pinturas formation, Early-Middle Miocene of Patagonia, biostratigraphic and paleoenvironmental implications. J South Am Earth Sci 18:199–212

Linares E, Llambías EJ, Latorre CO (1980) Geología de la provincia de La Pampa, República Argentina y geocronología de sus rocas metamórficas y eruptivas. Rev Asoc Geol Arg 35:87–146

Marquardt C, Blanco N, Godoy E, Lavenu A, Ortlieb L, Marchant M, Guzmán N (2000) Estratigrafía del Cenozoico Superior en el área de Caldera (26°45′-28°S). Cong Geol Chil 9:504–508

Marshall LG, Patterson B (1981) Geology and geochronology of the mammal-bearing Tertiary of the valle de Santa María and río Corral Quemado Catamarca province, Argentina. Fieldiana Geol 9:1–80

Martin L (1983) The origin and early radiation of birds. In: Brusch AH, Clark GA (eds) Perspectives in ornithology, essays presented for the centennial of the American Ornithologist Union. Cambridge University Press, New York

Mayr G (2005) "Old World phorusrhacids" (Aves: Phorusrhacidae): a new look at *Strigogyps* ("*Aenigmavis*") *sapea* (Peters 1987). PaleoBios 25:11–16

Mayr G (2009) Paleogene fossil birds. Springer, Heidelberg

Moreno FP, Mercerat A (1891) Catálogo de los pájaros fósiles de la República Argentina conservados en el Museo de La Plata. An Mus La Plata 1:7–71

Mourer Chauviré C (1981) Première indication de la présence de Phorusrhacides, famille d'oiseaux géants d'Amerique du Sul, dans le Tertiaire européen: Ameghinornis nov.ge, (Aves, Ralliformes) des Phosphorites du Quercy, France. Geobios 14:637–647

Nasif NL, Esteban GI, Ortiz PE (2009) Novedoso hallazgo de egagrópilas en el Mioceno tardío, Formación Andalhualá, provincia de Catamarca, Argentina. Ser Corr Geol 25:105–114

Nasif N (1988) Primer registro de flamencos (Phoenicopteridae) del Terciario Superior, Valle del Cajón (provincia de Catamarca, Argentina). Ameghiniana 25:169-173

Noriega JI (1992) Un nuevo genero de Anhingidae (Aves: Pelecaniformes) de la Formación Ituzaingó (Mioceno superior) de Argentina. Notas Mus La Plata 109:217–223

Noriega JI (1994) Las Aves del "Mesopotamiense" de la provincia de Entre Ríos Argentina. Dissertation, Universidad Nacional de La Plata

Noriega JI (1995) The avifauna from the "Mesopotamian" (Ituzaingó Formation: Upper Miocene) of Entre Ríos Province, Argentina. Cour Forsch Senck 181:141–148

Noriega JI (2000) Nuevos restos de Phororhacidae (Aves: Gruiformes) del "Mesopotamiense" (Fm. Ituzaingó; Mioceno tardío) en la provincia de Entre Ríos, Argentina. Ameghiniana 37:31R

Noriega JI (2001) Body mass estimation and locomotion of the Miocene pelecaniform bird *Macranhinga*. Acta Paleontol Pol 46:115–128

Noriega JI, Chiappe L (1993) An Early Miocene passeriform from Argentina. Auk 110:936–938

Noriega JI, Alvarenga HMF (2000) Phylogeny of the tertiary giant darters (Pelecaniformes: Ameghinidae) from South América. In: Proceedings of the 5th symposium of the society of avian paleontology and evolution, Beijing

Noriega JI, Piña CI (2004) Nuevo material de *Macranhinga paranensis* (Aves: Pelecaniformes: Anhingidae) del Mioceno Superior de la Formación Ituzaingó, provincia de Entre Ríos, Argentina. Ameghiniana 41:115–118

Noriega JI, Agnolín FL (2008) El registro paleontológico de las aves del "Mesopotamiense" (Formación Ituzaingó; Mioceno Tardío-Plioceno) de la provincia de Entre Ríos, Argentina. Insugeo 17:271–290

Noriega JI, Cladera G (2008) First record of an extinct marabou stork in the Neogene of South America. Acta Palaeontol Pol 53:593–600

Noriega JI, Vizcaíno SF, Bargo MS (2009) First record and a new species of seriema (Aves: Ralliformes: Cariamidae) from Santacrucian (Early-Middle Miocene) beds of Patagonia. J Vert Pal 29:620–626

Noriega JI, Areta JI, Vizcaíno SF, Bargo MS (2011) Phylogeny and taxonomy of the Patagonian Miocene Falcon *Thegornis musculosus* Ameghino, 1895 (Aves: Falconidae). J Paleontol 85:1089–1104

Olson S (1985) The fossil record of birds. In: Farner D, King J, Parkes K (eds) Avian biology. Academic Press, New York

Patterson B, Kraglievich L (1960) Sistemática y nomenclatura de las aves fororracoideas del Plioceno Argentino. Pub Mus Mun Cienc Nat Trad Mar del Plata 1:1–51

Picasso MBJ, Degrange FJ (2009) El género *Nothura* (Aves, Tinamidae) en el Pleistoceno (Formación Ensenada) de la provincia de Buenos Aires, Argentina. Rev Mex Cienc Geol 26:428–432

Picasso M, Tambussi CP, Dozo T (2009) Neurocranial and brain anatomy of a late Miocene eagle (Aves, Accipitridae) from Patagonia. J Vert Pal 29:1–6

Rasmussen DT, Kay RF (1992) A Miocene anhinga from Colombia, and comments on the zoogeographic relationships of South America's Tertiary avifauna. Sci Ser 36:225–230

Rinderknecht A, Noriega JI (2002) Un nuevo género de Anhingidae (Aves: Pelecaniformes) de la Formación San José (Plioceno-Pleistoceno) del Uruguay. Ameghiniana 39:183–192

Rovereto C (1914) Los estratos araucanos y sus fósiles. An Mus Nac Hist Nat Buenos Aires 25:1–247

Schultz PH, Zarate MA, Hames W, Camilión C, King J (1998) A 3,3—Ma impact in Argentina and possible consequences. Science 282:2061–2063

Sinclair W, Farr M (1932) Aves of the Santa Cruz beds. In: Scott W (ed) Reports of the Princeton University expeditions to Patagonia (1896–1899), vol 7. Princeton Univesity, New Jersey

Stucchi M (2003) Los Piqueros (Aves: Sulidae) de la formación Pisco, Perú. Bol Soc Geol Perú 95:75–91

Stucchi M (2008) Un condor del Mioceno tardío de la costa peruana. Bol Lima 153:141–146

Stucchi M, Emslie SD (2005) Un Nuevo Cóndor (Ciconiiformes, Vulturidae) del Mioceno Tardío-Plioceno temprano de la Formación Pisco, Perú. Condor 107:107–113

Sustaita D (2008) Musculoskeletal underpinnings to differences in killing behavior between North American accipiters (Falconiformes: Accipitridae) and falcons (Falconidae). J Morphol 269:283–301

Tambussi CP (1989) Las aves del Plioceno tardío-Plesitoceno temprano de la Provincia de Buenos Aires. Dissertation, Universidad Nacional de La Plata

Tambussi CP, Degrange FJ (2011) Las aves de Florentino Ameghino: estatus revisado. Ameghiniana 48:58R

Tambussi CP, Agnolin F, Cozzuol M (2003) Un nuevo predator en el elenco de aves de los ecosistemas miocénicos patagónicos. Ameghiniana 40:72R

Tambussi CP (1995) The fossil Rheiformes from Argentina. Cour Forch Senck 181:121–129

Tambussi CP (1997) Algunos aspectos biomecánicos de la locomoción de los fororracos (Aves, Gruiformes). Ameghiniana 34:541

Tambussi CP (2011) Paleoenvironmental and faunal inferences based upon the avian fossil record of Patagonia and Pampa: what works and what does not. Biol J Linn Soc 103:458–474

Tambussi CP, Noriega JI (1996) Summary of the Avian fossil record from southern South America. In: Arratia G (ed) Contributions of the southern South America to vertebrate paleontology. Müncher Geowissenschaftliche Abhandlungen

Tambussi CP, Noriega JI (1999) The fossil record of condors (Aves, Vulturidae) of Argentina. Smith Cont Pal 89:171–184

Tauber AL (1997a) Bioestratigrafía de la Formación Santa Cruz (Mioceno inferior) en el extremo sudeste de la Patagonia. Ameghiniana 34:413–426

Tauber AL (1997b) Paleoecología de la Formación Santa Cruz (Mioceno inferior) en el extremo sudeste de la Patagonia. Ameghiniana 34:517–529

Tonni EP (1974) Un nuevo cariámido (Aves, Gruiformes) del Plioceno Superior de la provincia de Buenos Aires. Ameghiniana 9:366–372

Tonni EP (1977) El rol ecológico de algunas aves fororracoideas. Ameghiniana 14:316

Tonni EP (1980) The present state of knowledge of the Cenozoic birds of Argentina. Contrib Sci 330:104–114

Tonni EP, Tambussi CP (1988) Un nuevo Psilopterinae (Aves: Ralliformes) del Mioceno tardío de la provincia de Buenos Aires, República Argentina. Ameghiniana 25:155–160

Tonni EP, Noriega JI (1998) Los cóndores (Ciconiformes, Vulturidae) de la región pampeana de la Argentina durante el Cenozoico tardío: distribución, interacciones y extinciones. Ameghiniana 35:141–150

Tonni EP, Noriega JI (2001) Una especie extinta de *Pseudoseisura* Reichenbach 1853 (Passeriformes: Furnariidae) del Pleistoceno de la Argentina: comentarios filogenéticos. Ornit Neot 12:29–44

Urbina M, Stucchi M (2005a) Evidence of a fossil stork (aves: ciconiidae) from the Late Miocene of the Pisco formation, Perú. Bol Soc Geol Perú 100:63–66

Urbina M, Stucchi M (2005b) Los cormoranes (Aves: Phalacrocoracidae) del Mio-Plioceno de la Formacion Pisco, Perú. Bol Soc Geol Perú 99:41–49

Vezzosi RI (2012) First record of *Procariama simplex* Rovereto, 1914 (Phorusrhacidae, Psilopterinae) in the Cerro Azul Formation (upper Miocene) of La Pampa Province; remarks on its anatomy, palaeogeography and chronological range. Alcheringa 1–13

Vignati MA (1925) La geología de Monte Hermoso. Physis VIII:126–127

Vizcaíno SF, Bargo MS, Kay RF, Milne N (2006) The armadillos (Mammalia, Xenarthra, Dasypodidae) of the Santa Cruz formation (early-middle Miocene): an approach to their paleobiology. Palaeogeog Palaeoecol 237:255–269

Vizcaíno SF, Bargo MS, Kay RF, Fariña RA, Di Giacomo M, Perry JMG, Prevosti FJ, Toledo N, Cassini GH, Fernicola JC (2010) A baseline paleoecological study for the Santa Cruz formation (late-early Miocene) at the Atlantic coast of Patagonia, Argentina. Palaeogeog Palaeoecol 292:507–519

Walsh SA (2002) Taphonomy and genesis of the Neogene Bahía Inglesa formation bonebed, northern Chile. J Vert Pal 22:117A

Walsh SA, Hume JP (2001) A new Neogene marine avian assemblage from north-central Chile. J Vert Pal 21:484–491

Walsh SA, Naish D (2002) Fossil seals from late Neogene deposits in South America: a new pinniped (Carnivora, Mammalia) assemblage from Chile. Palaeontol 45:821–842

Chapter 8
The Dominance of Zoophagous Birds: Just a Cliché?

It has been suggested that in South American ecosystems during Cenozoic times, carnivorous birds were hegemonic not only over any other trophic avian guild but also over other vertebrate carnivorous groups (Tambussi 2011 and the literature cited therein). To investigate this, we have chosen to dissect the avian fossil record to determine whether this assumption fits with the available data.

Before advancing, let us specify at the outset some concepts. In general, the diet of a zoophagous bird is based on animals (e.g., feed on animal matter) and includes avivores, mamalivores, herpetivores, piscivores, insectivores, scavengers, and even generalists (Hertel 1995). In this sense, zoophagous is a more comprehensive term than carnivorous, while the latter only means "meat-eaters". Here, we prefer to use zoophagous instead of the classic carnivorous term.

Most living birds (about 8,600 species) are zoophagous and roughly 60 % are partly or largely insectivorous (Morse 1971). It is generally known that bird's diets are not restricted to a single item because the trophic categories are arbitrary. Moreover, many birds change their diet throughout the year depending on food availability. Because insects significantly react to temperature or moisture changes, insectivorous birds will be affected. Migration is one of the possible responses when insects are not available in abundance. The remaining birds could partially change their diets to intake other alternatives (Morse 1971). Conversely, larger predaceous birds (e.g., eagles, owls) feed mainly on homeothermic prey that may not conspicuously respond to climatic changes (Morse 1971), and therefore it is expected that these birds have a more homogeneous diet throughout the year.

Another aspect about zoophagy is how it acquires and processes the food. Altogether, these activities comprise what is known as feeding strategies meaning "the set of choices made by the predator on encountering food items to eat an item or ignore it" (Orians 1971; Pulliam 1974, p. 3; Ferry-Graham et al. 2002). Predaceous zoophagous or predators run, fly, or swim to hunt and kill other animals to feed. It is obviously a strenuous activity made only possible by the presence of appropriate musculoskeletal structures and fitting senses. A predator should be able

C. P. Tambussi and F. J. Degrange, *South American and Antarctic Continental Cenozoic Birds*,   87
SpringerBriefs in Earth System Sciences, DOI: 10.1007/978-94-007-5467-6_8,
© The Author(s) 2013

(theoretically) to assess potential prey vulnerability, minimizing variability in hunting success (Quinn and Cresswell 2004). We will refer later to this topic. Seriemas, diurnal and nocturnal raptors, and as discussed below, also phorusrhacids are examples of predatory birds that chase their preys. A second type of predators include birds that wait for the prey in ambush, i.e., expect quietly in order to catch and attack or kill the prey. Whereas the first type of predator spends much time and energy in pursuit, the second one does it in searching the prey (Schoener 1971). These are type I predator and type II predators, respectively, of Schoener (1969). Examples of the latter are storks and herons that are commonly associated with aquatic environments. Some other predators feed within the water. This is the case for anhingids and cormorants that are in fact type I predators and we call them aquatic zoophagous. Predation has a strong influence upon animal populations and it represents a selective force in the evolution of form and function (Newton 1998; Caro 2005) either for the prey or the predator. It is evident that in each case the time and energy utilized to obtain the food and the amount of energy gained per capture will be different. Nonetheless, it is obvious that birds display a wide variety of feeding specializations and the aforementioned classification is unrealistic. However, it will be useful for the discussion that we will face here.

Ecosystems can be evaluated in at least two ways, taxonomic (answering the question what is it?) or functional (what does it do?). The first approach is better to evaluate biodiversity; the second one should be the most useful if the goal is to characterize the ecosystem condition (Cummins et al. 2005; Farlow and Pianka 2002). We are particularly interested in the latter. Functional approach has been applied since the 1970s (Cummins and Wilzbach 1985; Merritt and Cummings 1966), and is based on easily recognized morphological or behavioral features of the animals that are related to their modes of food acquisition. Needless to say, to describe the ideal structure of a given ecosystem, i.e., build an idea of how it works, allow us to predict the discovery of missing links or evaluate potential overlaps or separations of niche.

To analyze past ecosystems no other options but to accept certain licenses must be embraced. In most cases, it is not possible to state the absolute synchrony (simultaneous occurrence), coexistence or syntopy (co-occurrence in the same macrohabitat at the same locality during the same time) of individuals/species (Rivas 1964) that are under scrutiny. In this study, we jointly analyze fossils that came from the same site or sites separated by only a few tens or hundreds of kilometers without fear of implausible results. However, temporal differences or gaps are difficult to discriminate and could perhaps result in overlapping fauna that was not. Here, we use "associations" to gather records that come from near paleontological sites of the same geological formation (i.e., of similar age). Cerro Azul or Santa Cruz Formations are such cases.

Here, we will focus our analysis on zoophagous birds, understanding that this means that we are making a rough cutout to analyze the structure of the ecosystem. The more accurate analogy that comes to our mind is that we are watching only one picture of a full-length movie in which all the players were not on the scene. Therefore, the reader may feel that restriction of this discussion has resulted in a rather limited view of a much larger phenomenon when we are talking about ecosystem structure.

We have compared the zoophagous versus nonzoophagous avian record of 13 fossil localities of South America (four Paleogene and the remaining Miocene or Pliocene in age) presenting the results in summary in Table 8.1. Birds from each of these locations were separated according to their inferred diet, grouping in different columns zoophagous and nonzoophagous birds. Analyzing the ratio between these categories and the total record of each locality, one can realize that in all but one, proportion of zoophagous is higher and values are greater than 60 %. For the Late Oligocene Deseadan age, we considered six records, three of which (50 %) are zoophagous. However, we have highlighted the incompleteness and inaccuracy of their exact provenances, so these results are questionable.

Remember now that about 60 % of the living birds are partly or largely insectivorous and the percentage will be higher if the remaining zoophagous birds are taken together. Extrapolating this to the past, we could hypothesize that future discoveries will correspond roughly to nonzoophagous bird species and very few will be zoophagous. These assumptions are unrefined in many aspects. First, the fossil record is clearly biased toward medium- to big-sized birds; meanwhile small to very small birds are scarce (e.g., Passeriformes, Piciformes) surely due to the collecting style, preservation, and other taphonomic aspects of birds in particular.

The compared percentages were calculated taking all the birds without discriminating each type of environment. The biomass of each trophic guild (zoophagous versus nonzoophagous) is not considered, neither their home range. The latter should be greater in animals which are efficient pursuers than less efficient pursuers. These limitations are important in the analysis of the ecosystem structure.

Returning to Table 8.1, birds from Tremembé Formation are one particular case: the only three species of zoophagous birds registered for this formation (*Paraphysornis*, *Brasilogyps*, and *Taubatornis*) have been pointed out as scavengers. Meanwhile, in the other compared localities, the nonscavenger zoophagous birds are the dominant ones and only in five of these localities, there is at least one scavenger. The role assignment of *Paraphysornis brasiliensis* as a scavenger has been based on the morphology of the tarsometatarsus (short and wide) and a supposedly slow motion (Alvarenga and Höfling 2003). In the absence of the pelvis, Degrange (2012) points out that there is no reliable assignment of the locomotor habit. Furthermore, the cranial remains recovered of *Paraphysornis* include the jaw and a quadrate bone whose morphology on its own is not indicative of trophic habit. In this sense, the assignment of trophic habit should be taken with caution.

The estimated body mass for this phorusrhacid is 180 kg (Alvarenga and Höfling 2003), and in this sense is one of the five South American fossil zoophagous birds whose estimated body masses exceeds 100 kg: the Oligocene *Physornis fortis* (body mass unknown), the Patagonian Miocene gigantic predators *Phorusrhacos longissimus* (120 kg, Degrange 2012) and *Kelenken guillermoi* (skull of 716 mm long, unestimated body mass), and the Pliocene *Devincenzia pozzi* (161 kg, Degrange 2012). In any locality is verified the simultaneous coexisting of such gigantic zoophagous birds. This could be reflecting a very well-known situation which is that one of the ways that animals are thought to coexist is

Table 8.1 Main fossil localities and zoophagous versus nonzoophagous proportion

Procedence	Age	Birds		Zoophagous proportion (%)
		Zoophagous	Nonzoophagous	
Itaboraí SALMA	Late Paleocene	*Paleopsilopterus Itaboravis*	*Diogenornis*	66.6
La Meseta formation	Middle Eocene– Late Eocene	Pelagornithidae Procellariiformes Charadriiformes Gaviidae Polyborinae	Phoenicopteriformes Rallidae Ratitae	62.5
Deseadense SALMA	Late Oligocene	*Andrewsornis Psilopterus Physornis*	*Teleornis Aminornis Loxornis*	50
Tremembé formation	Late Oligocene– Early Miocene	*Paraphysornis Brasilogyps Taubatornis*	*Hoazinavis Chaunoides Agnopterus Ameripodius Palaelodus Taubacrex*	33.3
Santa Cruz formation	Late Early Miocene	*Patagornis Phorusrhacos Psilopterus* (2 spp.) *Thegornis* (2 spp.) *Badiostes* "Gruiformes" *Cariama* Cariaminae indet. *Liptornis Protibis*	*Opisthodactylus* Tinamidae (2 spp.) *Eoneornis Eutelornis Ankonetta Brontornis*	63.15
Puerto Madryn formation	Early Miocene	*Leptoptilus Geranoaetus* Psilopterinae Patagornithinae	Dendrocygninae	80
Pinturas formation	Early Miocene	Polyborinae Strigidae Cariamidae Tyranni	Tinamidae Anatidae	66.6
Cerro Azul formation	Late Miocene	*Milvago* Tyrannidae *Argentavis Procariama* Patagornithinae indet.	*Eudromia Nothura Pterocnemia*	62.5

(continued)

Table 8.1 (continued)

Procedence	Age	Birds		Zoophagous proportion (%)
		Zoophagous	Nonzoophagous	
Ituzaingó formation	Late Miocene– Early Pliocene	*Macranhinga* (2 spp.) *Anhinga* Cf. *Giganhinga* Mycterini ˙Ciconini *Grus* *Andalgalornis* *Devincenzia* Phorusrhacinae indet.	Phoenicopteridae *Megapalaelodus* Rallidae Dendrocheninae Anatini Rheidae	62.5
Andalhualá formation	Late Miocene– Early Pliocene	*Andalgalornis* *Procariama* *Mesembriornis* *Geranoaetus*	Palaeolodidae indet. *Megapalaelodus*	66.6
Pisco formation	Late Miocene– Early Pliocene	Sulidae Lariidae Pelagornithidae Procellariidae Scolopacidae Pelecanidae Diomedeidae Phalacrocoracidae Vulturidae Ciconidae	–	100
Bahia Inglesa formation	Mio-Pliocene	Pelagornithidae Sulidae	–	100
Chapadmalal formation	Pliocene	*Mesembriornis* Mesembriornithinae nov. gen. et sp. *Vultur* Furnariidae *Charadrius* *Calidris*	*Hinasuri* Tinamidae (2 spp.)	66.6

SALMA South American land mammal age

by differences in their body sizes, and this is because different sized animals utilize different resources (Wilson 1975). The exploitation of the same resource of species inhabiting a particular community may cause interspecific competition (endeavor of two or more animals to gain the same particular thing in the sense of Milne 1961). Tentatively, it is said that there is a theoretical limit to morphological similarity among competing species, ideally manifested as a Hutchinson ratio of mean sizes among successive pairs of competing species: "where the species co-occur, the ratio of the larger to the small form varies 1.1–1.4, the mean ratio being roughly 1.3" (Hutchinson 1959, p. 152 but see Wiens 1982). For studies of

Table 8.2 Hutchinson's (1959) ratios calculated for the Santacrucian zoophagous birds

Taxa	Body mass (kg)[a]	Culmen length (cm)	Hutchinson's ratios[b]	
			Body mass	Culmen length
Phorusrhacos longissimus	93	–	3.57	–
Patagornis marshi	26	25.19	3.25	1.87
Psilopterus lemoinei	8	13.4	1.77	1.13
Psilopterus bachmanni	4.5	11.8	2.36	1.74
Thegornis musculosus	1.9	1.5	1.26	4.52
Cariama santacrucensis	1.5	6.78	1.04	–
Anisolornis excavatus	1.43	–	1.47	–
Liptornis hesternus	0.97	–	4.61	–
Badiostes patagonicus	0.21	–	–	–

[a] According to Degrange et al. (2012). Species were arranged according to a decreasing body mass

[b] Ecologists recognize 1.3 as a critical value of this ratio which commonly facilitates coexistence. Species are deemed to be too similar to coexist if the ratio is below this value (see Hutchinson 1959; Wiens 1982)

birds, bill length or body mass are relevant (Hespenheide 1971). We calculate Hutchinson ratios for the Santacrucian birds considering two variables, body mass (as indicative of body size) and culmen length (as a gross indicative of food indicator, although discussed by several workers), but the data obtained give no indication of regularly spaced size ratios (Table 8.2). As we discussed in Chap. 7, species of the Santacrucian zoophagous guild (63 %) include falconids (three species), phorusrhacids (four species), cariamids (one species), ciconiiforms (one species), anhingids (one species), and gruiform (one species). Likewise, we point out two aspects: (1) except in two cases (*Cariama santacrucensis* and *Anisolornis excavatus*), the estimated body masses do not overlap and (2) the culmen length is similar in both *Psilopterus* species. Interestingly, one has twice the mass of the other. Probably they, as other zoophagous birds, differ in their hunting methods (best runners will be more effective pursuers), diet (proportion of meat, bones, and invertebrate prey in the diet), and some cases in habitat. Indeed, morphological information can provide some evidences for interspecific competition taking into account that to establish this kind of relationship in the fossil record is problematic.

Table 8.1 also shows that there are birds with more than 10 kg of body mass in 8 of the 13 localities and in all the cases, they are represented only by phorusrhacids and rheids. Nowadays, in South America, zoophagous birds exceeding 10 kg are the minority; to name a few, the condors weigh between 10 and 12 kg and the albatross between 8 and 10 kg (Dunning 2008).

It is known that predators and prey interact in a wide diversity of ways (Newton 1998; Krause and Ruxton 2002; Caro 2005). To predict the outcomes of predator–prey interactions seems to be really difficult in extant species (Quinn et al. 2008). So, what can we say about the possible interaction between predators that co-habitat in the same space? It seems that the biggest zoophagous have a strong dominance on the smaller zoophagous. However, we left for the future to explore this very attractive idea, i.e., the possible dominance of some zoophages over others in each particular environment. For this, it will be necessary to deepen the analysis studying the ecological framework, although it is known that the feeding relationships are usually complex and must be taken into account, if we have general theories of the structure of animal communities (Vezina 1985).

Few are the localities where there is a mixture of zoophagous birds. Only in 4 of the 13 localities (Santa Cruz, Chapadmalal, Ituzaingó, and Puerto Madryn) volant, aquatic, and terrestrial zoophagous are present. In these cases, the zoophagous diversity may reflect the presence of diverse environments. Meanwhile, in three localities (the marine La Meseta, Pisco, and Bahía Inglesa Formations), the aquatic zoophagous are dominant. In the other extreme, the Andalhualá Formation is basically dominated by terrestrial zoophagous birds (the terror birds).

Any study that intends to establish the trophic relationships or any other ecological frame requires paleoecological, morphological, and functional studies in-depth. Much progress has been made in this direction but still some studies are lacking, such as depth studies in which the record permits and in other cases where it is not possible until the discovery of new fossils.

Independently of the paleoecological relationships that can be established in the future, within the South American zoophagous birds very striking bird lineages outstand: teratorns and phorusrhacids. The latter especially has a set of morphological characteristics that have not been developed in any other group of birds.

Next, we comment about these two fantastic birds' lineages and other typical zoophagous birds that share or compete with (falcons, hawks, eagles, and vultures).

8.1 Teratorns

Teratornithidae constitute an extinct lineage of volant birds from big to giant body size, related with the Ciconiiformes and in particular with the Cathartidae (Olson 1985; Mayr 2009). However, there is still controversy about their affinities (Suárez and Olson 2009; Cenizo et al. 2011). Their biggest representation and diversity are recorded in the Pleistocene of North America (Florida, Arizona, Nevada, México, and California), although some remains have been recorded in Cuba, Perú, Brazil, and Argentina. The fossil record supports an origin in South America concomitant with the development of open environments (Campbell and Tonni 1980, 1981, 1983). It is plausible that teratorns reached North America by the Pleistocene (or Late Pliocene).

Six genera of Teratornithidae have been described: *Teratornis*, *Cathartornis*, and *Aiolornis* from USA, *Oscaravis* from Cuba, *Taubatornis* from Brazil, and *Argentavis* from Argentina.

Oscaravis olsoni from the Pleistocene of Cuba is the only insular member of this family and points out that the Teratornithidae were able to disperse crossing oceanic barriers (Olson and Alvarenga 2002). According to this, it can be hypothesized that the dispersion to North America was not conditioned by the presence of a land bridge and the dispersion could occurr anterior to the definitive Panamanian isthmus (circa 3 million years ago, Suárez and Olson 2009).

Taubatornis campbelli based on the distal portion of a left ulna and tibiotarsus (Olson and Alvarenga 2002) is a small Teratornithidae and corresponds to the oldest record of the family (Late Oligocene-Early Miocene).

Without doubt, the most astonishing teratorn is *Argentavis magnifiscens* whose remains have been recovered from the Miocene of the Cerro Azul Formation in La Pampa Province (Campbell and Tonni 1980; Cenizo et al. 2011) and in the Pliocene of the Andalhualá Formation from Catamarca Province, both from Argentina (Fig. 7.5). It has been suggested several times that *Argentavis* was the biggest flying bird from all times (Campbell and Tonni 1980, 1981, 1983; Campbell 1995; Vizcaíno and Fariña 1999), whose body mass raised up about 72 kg (Chatterjee et al. 2007).

The Teratornithidae trophic habit and flying style have been subjects of debate in the scientific literature. The amazing abundance of specimens recovered in the asphaltic deposits of Rancho La Brea in California (USA) allowed to suggest a scavenger habit taking advantage of the corpses that were entangled in the tar (Miller 1925; Fisher 1945; Howard 1962). A different proposal, although poorly accepted, was sustained by Campbell and Tonni (1981, 1983) who suggested that based on the quadrate mobility and the mandible opening, these birds were predators with the possibility to swallow their whole preys. Hertel (1995) proposed a piscivorous habit and eventually scavenger for Teratronis. Almost two decades after its original description (Campbell and Tonni 1980), estimating the home-range amplitude and the metabolic rate of *Argentavis magnificens*, the scavenger hypothesis is regaining strength (Vizcaíno and Fariña 1999), at least for the giant teratorn.

Although it is widely accepted the soaring flying style for these birds, its take-off capabilities are still controversial (running or jumping from heights (Campbell and Tonni 1983; Chatterjee et al. 2007; Vizcaíno and Fariña 1999).

8.2 Phorusrhacids

Phorusrhacids are large, land predator, nonaquatic, and poorly or nonvolant birds. These extinct Cariamiformes related to the extant seriemas (Andrews 1899; Alvarenga and Höfling 2003; Alvarenga et al. 2011; Agnolín 2009), are the most characteristic, diverse, and staring birds from the South American Cenozoic. With a very vast fossil record, these birds are distributed from the Eocene to the Pleistocene coming from Argentina (Degrange 2012), Brazil (Alvarenga et al. 2010), Uruguay (Montenegro et al. 2010; Tambussi et al. 1999), USA (Brodkorb 1963; Chandler 1994; Baskin 1995; MacFadden et al. 2007), and Africa (Mourer-Chauviré et al. 2011). Recently, some isolated remains from the middle Eocene of Europe have been reported as belonging to phorusrhacids (Angst and Bufeaut 2012a, b).

Phorusrhacidae include the Psilopterinae (small and gracile; Palaeocene to Pliocene), Mesembriornithinae (mid to big size, gracile-legged; Miocene to Pliocene), Patagornithinae (mid-sized; Oligocene to Pliocene), Phorusrhacinae (mostly large, gracile-legged; Miocene to Pleistocene), and Physornithinae (gigantic, robust; Oligocene). Due to the huge size of some of these birds, the term "terror birds" was raised in the popular and scientific literature. However, this term can produce confusion, because not all phorusrhacids are huge and therefore they are not "terror birds" in this sense (Degrange 2012).

In general, these birds are characterized by their elongated hindlimbs, narrow pelvis, reduced forelimbs, and their huge skull with their tall, long, narrow, and hollow beaks ended in a hook. In particular, the structure and design of the beak are exclusive features of the phorusrhacids, absent in any other avian, extinct or extant. Likewise, and contrary to the remainder Neornithes, phorusrhacids have lost their capacity to flexionate dorsoventrally their beak (a quality known as avian kinesis). This cranial immobility (akinesis) is unknown in other predator birds. The akinesis together with the shape of the beak (Degrange 2012; Degrange et al. 2010a, b) and the structure of the neck (Tambussi et al. 2012) are very appropriate to kill. Two recent studies deepen these aspects. A finite element and a biomechanical analysis applied to *Andalgalornis steulleti* suggests that the skull would be prepared to make sudden movements in the sagittal plane to subdue prey (Degrange et al. 2010a). The second one says that the musculoskeletal system of *Andalgalornis*' neck seems to be prepared to support a particularly big head during normal stance, and to help the neck (and the head) rising after the maximum ventroflexion during a strike (Tambussi et al. 2012). These are important starting points for inferring the behavior of these big headed birds with high and compressed beaks.

Some metric values calculated by Degrange (2012) using different techniques, clearly demonstrate that phorusrhacids were medium to giant birds. For example, body mass estimations extend from 4.5 (*Psilopterus bachmani*) to 120 kg or more (*Phorusrhacos longissimus*). For those in which it was possible to estimate the height, the values range from 88 cm (*Procariama simplex*), 162 cm (*Mesembriornis milneedwarsi*), or 125 cm (*Andalgalornis steulleti*). There is no, and nor was, any other place in the world where terrestrial predatory birds have evolved to this size.

Phorusrhacids dominated the terrestrial environment throughout the Cenozoic, as emphasized by the unusually large body size of some species as well as their numerical predominance with respect to other birds. They consist of several terrestrial species and are found almost exclusively in open habitats. Landscape heterogeneity may be important for these birds not only at regional but also at microscale, that is, the refinement in the environmental reconstruction is paramount. In other words, habitat heterogeneity had been important for phorusrhacid species richness in South America. It is not a trivial matter to analyze this problem in the future, but is not the approach here.

Phorusrhacids decline in diversity toward the end of the Pliocene and their last records in South America date from the Pleistocene (Tambussi et al. 1999; Alvarenga et al. 2010; Montenegro et al. 2010). There are two explanatory hypotheses proposed for this decline, either for environmental reasons or direct competition (at least for the larger phorusrhacids) with placental carnivore's immigrants to South America after the setting of the Panamanian bridge. Especially, the second hypothesis requires deep analysis of the trophic relationships established in South America before and after the Great American Biotic Interchange.

8.3 Falcons, Eagles, Condors, Vultures, and Owls

Together with Phorusrhacids and Teratorns, few other taxa formed the set of South American Cenozoic zoophagous birds.

Falcons (Falconidae) represent a largely distributed lineage of raptorial birds, but unfortunately very poorly represented in the fossil record. They show diverse behaviors, encompassing both the aerial swift hunting falcons and the neotropical generalists and carrion-eating caracaras. In falcons, their strategy of capture generally consists of striking their prey in flight, but they kill them by the action of the beak on the neck, breaking the spinal vertebrae and damaging the medulla (Sustaita 2008).

In the case of the Polyborinae caracaras (=Caracarinae), the trophic habits are clearly scavengers and the morphology of their hindlimbs is quite different from the rest of the falcons, with long tarsometatarsus. They represent a morphological conservative taxon since their first records in the Eocene of the Antarctica.

Eagles and Hawks (Accipitridae) are zoophagous birds from small to big size, with a high degree of morphological disparity, distributed all over the world, except Antarctica. Accipitridae are characterized by the conspicuous beak hook and their powerful talons equipped with strongly curveted claws and by a wide variation in wing shape according to the environment where they live (Thiollay 1994). They tend to suffocate their prey or to perforate the prey's lungs using their mighty talons (Sustaita 2008). As in the case of the Falconidae, their fossil record is very scarce.

Traditionally, the Cathartidae or "new world vultures" have been related with the diurnal raptors (Accipitridae and Falconidae). However, since a few decades, based on molecular studies, they have been placed within the Ciconiiformes (Sibley and Ahlquist 1990). Cathartidae are characteristic birds by their soaring flying style and their feeding based on carrion (Hertel 1994). Feathers of the head and neck are absent, the beak has a sharp cutting edge and is distally curved. Usually, they feed stepping on their prey and pull out the rotten meat through strong pullbacks (Houston 1994).

While its current distribution is American, there are some fossil records of Cathartidae from the Eocene and Oligocene of France (Oberholser 1899; Gaillard 1908), which indicate that their actual distribution may be relictual (Mayr 2005). Contrary to falcons and eagles, the fossil record of Cathartidae is quite abundant. The oldest record comes from the Middle to Late Eocene of France (Mourer-Chauviré 2002). In North America, the oldest records correspond to the species *Palaeogyps prodromus* and *Phasmagyps patritus* form the Oligocene of Colorado (Wetmore 1927). As the Teratornithidae, the Cathartidae achieve a big diversity in the Pleistocene of USA, including vultures (*Breagyps*, *Cathartes*, and *Coragyps*) and condors (*Hadrogyps*, *Pliogyps*, *Aizenogyps*, *Geronogyps*, and *Gymnogyps*) indicating that both forms were equally diverse.

Unlike North America and with the exception of *Brasilogyps*, the South American fossil record is fundamentally represented by condors (Tambussi and Noriega 1996; Alvarenga et al. 2008). It is worthy to mention here that the oldest condors from South America are *Perugyps diazi* from the Middle Miocene to Early Pliocene of Perú, meanwhile *Dryornis pampeanus* and *Vultur gryphus*, both from the Early to Middle Pliocene (Tonni and Noriega 1998; Tambussi and Noriega 1999) are the oldest records for condors in the Argentinean Pampean region.

Pleistocene birds will not receive further attention in this work, but the fossil record of condors during this epoch is so vast that it is worthy to mention it here. In Peru, the condors *Geronogyps reliquus*, *Gymnogyps howardae*, *Sarcoramphus fisheri*, *Vultur*, *Coragyps*, and *Cathartes* have been recorded and they are all genera with extant representatives and the last three possibly belong to the extant species (Campbell 1979; Tambussi and Noriega 1999; Stucchi and Emslie 2005; Alvarenga et al. 2008).

Geronogyps reliquus is the only condor species of the extinct genus *Geronogyps*, originally described for the Late Pleistocene of Peru and discovered more than 3,000 km to the southeast in the Pleistocene *s.l.* of Buenos Aires and Late Pleistocene of Entre Ríos in Argentina (Tonni and Noriega 1998; Tambussi and Noriega 1999; Noriega and Tonni 2007). Also, for the Pleistocene of the Pampean region *Sarcoramphus papa* is reported more than 700 km south to the most austral limit of the actual distribution of this taxon (Noriega et al. 2002; Noriega and Areta 2005).

It is registered in Bolivia *Sarcoramphus patruus*, and in Brazil the species *Pleistovultur nevesi*, *Vultur gryphus*, *and Wingegyps cartellei* and one indeterminate species of Cathartidae from the Late Pleistocene to Early Holocene (Alvarenga 1998; Alvarenga and Olson 2004; Alvarenga et al. 2008).

Wingegyps cartellei is a small enigmatic condor with characters remarkably similar to the living California condor *Gymnogyps* (Alvarenga and Olson 2004). During the Pleistocene, *Gymnogyps* was widespread across the Americas.

We highlight here the findings in Cuba of *Sarcoramphus* sp. and *Gymnogyps varonai* which allows to infer dispersion through maritime ways (Tambussi and Noriega 1999; Noriega and Areta 2005).

The wide diversity of Cathartidae during the South American Pleistocene is probably related with the diversity of the megafauna and with a niche distribution similar to that observed in the old world vultures (Accipitridae) (Alvarenga et al. 2008). The diminishing in the species richness would occur toward the end of the Pleistocene, concomitantly with the development of wide forested areas and the retraction of the megafauna (Tonni and Noriega 1998; Alvarenga et al. 2008). In South America, the availability of carcasses of these huge mammals could consist in the main source of food for these scavenger birds (Emslie 1987; Tonni and Noriega 1998; Alvarenga et al. 2008).

The earliest representatives of the owls (Strigiformes), a group of birds that typically hunt mammals during the night, are from the Late Paleocene of North

America and France. Characteristic pellets containing remains of small mammals occur at a Late Paleocene locality in China. The fossil record of Strigiformes is one of the most extensive among the neornithine birds, but restricted geographically to Europe and North America (Kurochkin and Dyke 2011). Brodkorb (1971) listed some 41 extinct species of owls, 11 of which were Tytonidae, 5 were Protostrigidae, and 25 were Strigidae. In South America, the only record for a Strigidae comes from the Pinturas Formation (Early Miocene).

8.4 A Holistic View: The Zoophagous Bird Landscape

Viewing all again, a dramatic turnover occurred after the end of the Cretaceous, and mammals were facing a completely different scenario. At the beginning of the Paleocene, terrestrial carnivores included crocodiles, snakes, and lizards that were potential predators and could have hunted mammals or other reptiles, including birds (Tambussi 2011). But if something is known of South America it is the variety of nonmammalian carnivores that dominated there. This is a different situation from the rest of the world in which both the carnivore and the herbivore role is dominated by placental mammals. So, a peculiar feature of earliest Cenozoic ecosystems is the absence of large placental mammal predators.

More specifically, the South American vertebrate Cenozoic record is well known for its impressive assemblage of zoophagous birds. Of particular interest are the diversity and large number of phorusrhacid specimens. Within the Phorusrhacidae, we see the first birds to occupy an ecological niche as large cursorial predators in South America.

During the Paleocene to Pliocene times, the extinct carnivorous Sparassodonta (Hathliacynidae and Borhyaenidae) also evolved in South America occupying the adaptive zone for large mammalian predators of terrestrial ecosystems. This group of metatherian mammals includes different morphotypes ranging from opossum to bear sizes. Tooth architecture was purely carnivore and some taxa were bone-eaters (Forasiepi et al. 2007). Large flightless birds and large sparassodonts possibly have competed for some identical resources in South America. Other members of the zoophagous bird fauna include falcons, accipitrids, storks and allies, anhingids, condors, vulture-like condors, owls, and teratorns.

References

Agnolín FL (2009b) Sistemática y filogenia de las aves fororracoideas (Gruiformes: Cariamae). Monografías Fundación Azara

Alvarenga HMF (1998) On the occurrence of the condor (*Vultur gryphus*) in the Holocene of the Lagoa Santa, Minas Gerais, Brazil region. Ararajuba 6:60–63

Alvarenga HMF (2010) *Diogenornis fragilis* Alvarenga, 1985, restudied: a South American ratite closely related to Casuariidae. In: 25th International Ornithology Congress, p 143

Alvarenga HMF, Höfling E (2003) Systematic revision of the Phorusrhacidae (Aves: Ralliformes). Pap Av Zool 43:55–91

Alvarenga HMF, Olson SL (2004) A new genus of tiny condor from the Pleistocene of Brazil (Aves: Vulturidae). Proc Biol Soc Wash 117:1–9

Alvarenga HMF, Guilherme RR, Brito RM, Hubbe A, Höfling E (2008) *Pleistovultur nevesi* gen. et sp. nov. (Aves: Vulturidae) and the diversity of condors and vultures in the South American Pleistocene. Ameghiniana 45:613–618

Alvarenga HMF, Chiappe LM, Bertelli S (2011) Phorusrhacids: the terror birds. In: Dyke G, Kaiser G (eds) Living dinosaurs: the evolutionary history of modern birds. Wiley, Chichester

Andrews C (1899) On the extinct birds of Patagonia, I, the Skull and skeleton of *Phororhacos inflatus* Ameghino. Trans Zool Soc Lond 15:55–86

Angst D, Buffetaut E (2012a) A large Phorusrhacid bird from the middle Eocene of France. In: Worthy TH, Göhlich UB (eds) 8th international meeting of society of avian paleontology and evolution, Abstracts

Angst D, Buffetaut E (2012b) A large phorusrhacid from the Middle Eocene of France and its palaeobiogeographical implications. 26 Jor Arg Pal Vert, Abstracts available in CD

Baskin JA (1995) The giant flightless bird *Titanis walleri* (Aves: Phorusrhacidae) from the Pleistocene coastal plain of south Texas. J Vertebr Paleontol 15:842–844

Brodkorb P (1963) A giant flightless bird from the Pleistocene of Florida. Auk 80:111–115

Brodkorb P (1971) Catalogue of fossil birds. Part IV (Columbiformes through Piciformes). Bull Fla State Mus Biol Sci 15:163–266

Campbell KE Jr (1979) The non-passerine Pleistocene avifauna of Talara Tar Seeps, northwestern Perú. Life Sci Contrib R Ontario Mus 118:1–203

Campbell KE Jr (1995) Additional specimens of the giant teratorn, *Argentavis magnificens*, from Argentina (Aves: Teratornithidae). C For Senck 181:199–201

Campbell KE Jr, Tonni EP (1980) A new genus of teratorn from the Huayquerian of Argentina (Aves: Teratornithidae). Contrib Sci 330:59–68

Campbell KE Jr, Tonni EP (1981) Preliminary observations on the paleobiology and evolution of teratorns (Aves, Teratornithidae). J Vertebr Paleontol 1:265–272

Campbell KE Jr, Tonni EP (1983) Size and locomotion in teratorns (Aves: Teratornithidae). Auk 100:390–403

Caro TM (2005) Antipredator defenses in birds and mammals. University of Chicago Press, Chicago

Cenizo MM, Tambussi CP, Montalvo CI (2011) Late Miocene continental birds from the Cerro Azul formation in the Pampean region (central-southern Argentina). Alcheringa 1–22. doi: 10.1080/03115518.2011.582806

Chandler RM (1994) The wing of *Titanis walleri* (Aves: Phorusrhacidae) from the late Blancan of Florida. Bull Fla Mus Nat Hist 36:175–180

Chatterjee S, Templin J, Campbell JK Jr (2007) The aerodynamics of *Argentavis*, the world's largest flying bird from the Miocene of Argentina. PNAS 104:12398–12403

Cummins KW, Wilzbach MA (1985) Field procedures for analysis of functional feeding groups of stream invertebrates. Appalachian Environmental Laboratory, University of Maryland, Maryland

Cummins K, Merrit R, Andrade P (2005) The use of invertebrate functional grouped to characterize ecosystem attributes in selected streams and rivers in south Brazil. Studies on Neotropical fauna andenvironments 40:69–89

Degrange FJ (2012) Morfología del cráneo y complejo apendicular posterior de aves fororracoideas: implicancias en la dieta y modo de vida. Dissertation, Universidad Nacional de La Plata

Degrange FJ, Tambussi CP, Moreno K, Witmer LM, Wroe S (2010a) Mechanical analysis of feeding behavior in the extinct "Terror Bird" *Andalgalornis steulleti* (Gruiformes: Phorusrhacidae). PLoS ONE. doi:10.1371/journal.pone.0011856

Degrange FJ, Tambussi CP, Jones WW, Blanco ER (2010b) Fororracos (Aves, Paleoceno-Pleistoceno): pérdida de quinesis craneana e implicancias funcionales. Actas X Cong Arg Pal Bioestrat VII Cong Lat Pal, pp 156–157

Degrange FJ, Noriega JI, Areta JI (2012) Diversity and paleobiology of the santacrucian birds. In: Vizcaíno SF, Kay RF, Bargo MS (eds) Early Miocene paleobiology in Patagonia: high-latitude paleocommunities of the Santa Cruz formation. Cambridge University Press, Cambridge

Dunning JB (2008) Handbook of avian body masses. CRC Press, Florida

Emslie SD (1987) Age and diet of fossil California condors in Grand Canyon, Arizona. Science 237:768–770

Farlow JO, Pianka ER (2002) Body size overlap, habitat partitioning and living space requirements of terrestrial vertebrate predators: implications for the paleoecology of large theropod dinosaurs. Hist Biol 16:21–40

Ferry-Graham L, Bolnick DI, Wainwright PC (2002) Using functional morphology to examine the ecology and evolution of specialization. Integr Comp Biol 42:265–277

Fisher HI (1945) Locomotion in the fossil vulture Teratornis. Am Midl Nat 33:725–742

Forasiepi A, Martinelli AG, Goin FJ (2007) Revisión taxonómica de *Parahyaenodon argentinus* Ameghino y sus implicancias en el conocimiento de los grandes mamíferos carnívoros del Mio-Plioceno de América de Sur. Ameghiniana 44:143–159

Gaillard C (1908) Les oiseaux des Phosphorites du Quercy. Ann Univ Lyon (Nouv Ser) 23:1–178

Hertel F (1994) Diversity in body size and feeding morphology within past and present vulture assemblages. Ecology 75:1074–1084

Hertel F (1995) Ecomorphological indicators of feeding behavior in recent and fossil raptors. Auk 112:890–903

Hespenheide HA (1971) Food preference and the extent of overlap in some insectivorous birds, with special reference to the Tyrannidae. Ibis 113:59–72

Houston DC (1994) Family Cathartidae (New World vultures). In: del Hoyo J, Elliott A, Sargatal J (eds) Handbook of the birds of the world. New World vultures to guineafowl, vol 2. Lynx Edicions, Barcelona

Howard H (1962) A comparison of prehistoric avian assemblages from individual pits at Rancho La Brea, California. Contrib Sci 58:1–24

Hutchinson GE (1959) Homage to Santa Rosalia, or why are there so many kinds of animals? Am Nat 93:145–159

Krause J, Ruxton GD (2002) Living in groups. Oxford University Press, Oxford

Kurochkin NE, Dyke GJ (2011) The first fossil owls (Aves: Strigiformes) from the Paleogene of Asia and a review of the fossil record of Strigiformes. Paleontol J 45:445–458

MacFadden BJ, Labs-Hochstein J, Hulbert RC, Baskin JA (2007) Revised age of the late Neogene terror bird (Titanis) in North America during the Great American Interchange. Geology 35:123–126

Mayr G (2005) "Old World phorusrhacids" (Aves: Phorusrhacidae): a new look at *Strigogyps* ("*Aenigmavis*") *sapea* (Peters 1987). PaleoBios 25:11–16

Mayr G (2009) Paleogene fossil birds. Springer-Verlag, Berlin

Merritt RW, Cummings KW (1966) Trophic relations of microinvertebrates. In: Hauer FR, Lamberti GA (eds) Methods in stream ecology. Academic Press, San Diego

Miller L (1925) The birds of Rancho La Brea. Carneg Inst Wash 349:63–106

Milne A (1961) Definition of competition among animals. Symp Soc Exp Biol 15:1–39

Montenegro F, Jones WW, Lecuona G, Toriño P, Batista A, García G, Ubilla D (2010) Nuevos aportes al conocimiento de los Phorusrhacinae (Aves, Phorusrhacidae) del Pleistoceno tardío. Res X Cong Arg Pal Bioestrat VII Cong Lat Pal, p 186

Morse DH (1971) The foraging of warblers isolated on small islands. Ecology 52:216–228

Mourer-Chauviré C (2002) Revision of the Cathartidae (Aves, Ciconiiformes) from the Middle Eocene to the Upper Oligocene Phosphorites du Quercy, France. In: Zhou Z, Zhang F (eds) Proceedings of 5th Symposium of Society of Avian Paleontology and Evolution. Science Press, Beijing

Mourer-Chauviré C, Tabuce R, Mahboubi M, Adaci M, Bensalah M (2011) A Phororhacoid bird from the Eocene of Africa. Naturwissenschaften. doi:10.1007/s00114-011-0829-5

Newton I (1998) Population limitation in birds. Academic Press, London

Noriega JI, Areta JI (2005) First record of *Sarcoramphus* Dumeril 1806 (Ciconiiformes: Vulturidae) from the Pleistocene of Buenos Aires province, Argentina. J South Am Earth Sci 20:73–79

Noriega JI, Tonni EP (2007) Geronogyps reliquus Campbell (Ciconiiformes: Vulturidae) in the Late Pleistocene in Entre Rios Province and its paleoenvironmental significance. Ameghiniana 44:245–250

Noriega JI, Areta JI, Dondas A (2002) Primer registro de *Sarcoramphus* Dumeril 1806 (Ciconiiformes: Vulturidae) en el Pleistoceno de la provincia de Buenos Aires. 8° Cong Arg Pal Bioestrat, p 50

Oberholser HC (1899) Some untenable names in ornithology. Proc Acad Nat Sci 51:201–216

Olson S (1985) The fossil record of birds. In: Farner D, King J, Parkes K (eds) Avian biology. Academic Press, New York

Olson SL, Alvarenga HMF (2002) A new genus of small teratorn from the Middle Tertiary of the Taubaté Basin, Brazil (Aves: Teratornithidae). Proc Biol Soc Wash 115:701–705

Orians G (1971) Ecological aspects of behavior. In: Farner D, King J, Parkes K (eds) Avian biology. Academic Press, New York

Pulliam HR (1974) On the theory of optimal diets. Am Nat 108:59–74

Quinn JL, Cresswell W (2004) Predator behaviour and prey vulnerability. J Anim Ecol 73:143–154

Quinn JL, Reynolds SJ, Bradbury RB (2008) Birds as predators and as prey. Ibis 15:1–8

Rivas LR (1964) A reinterpretation of the concepts 'sympatric' and 'allopatric' with proposal of the additional terms 'syntopic' and 'allotopic'. Syst Zool 13:42–43

Schoener TW (1969) Optimal size and specialization in constant and fluctuating environments: an energy-time approach. Brookhaven Symp Biol 22:103–114

Schoener TW (1971) Theory of feeding strategies. Ann Rev Ecol Syst 2:369–404

Sibley CG, Ahlquist JE (1990) Phylogeny and classification of birds: a study in molecular evolution. Yale University Press, New Haven

Stucchi M, Emslie SD (2005) Un Nuevo Cóndor (Ciconiiformes, Vulturidae) del Mioceno Tardío-Plioceno temprano de la Formación Pisco, Perú. Condor 107:107–113

Suárez W, Olson SL (2009) A new genus for the Cuban teratorn (Aves: Teratornithidae). Proc Biol Soc Wash 122:103–116

Sustaita D (2008) Musculoskeletal underpinnings to differences in killing behavior between North American accipiters (Falconiformes: Accipitridae) and falcons (Falconidae). J Morphol 269:283–301

Tambussi CP (2011) Paleo environmental and faunal inferences based upon the avian fossil record of Patagonia and Pampa: what works and what does not. Biol J Linn Soc 103:458–474

Tambussi CP, Noriega JI (1996) Summary of the avian fossil record from Southern South America. In: Arratia G (ed) Contributions of the southern south America to vertebrate paleontology. Müncher Geowissenschaftliche Abhandlungen

Tambussi CP, Noriega JI (1999) The fossil record of condors (Aves, Vulturidae) of Argentina. Smith Cont Pal 89:171–184

Tambussi CP, Ubilla M, Perea D (1999) The youngest large carnassial bird (Phorusrhacidae, Phorusrhacinae) from South America (Pliocene-Early Pleistocene of Uruguay). J Vertebr Paleontol 19:404–406

Tambussi CP, de Mendoza R, Degrange FJ, Picasso MBJ (2012) Flexibility along the neck of the neogene terror bird Andalgalornis steulleti (Aves Phorusrhacidae). PLoS ONE. doi:10.1371/journal.pone.0037701

Thiollay JM (1994) Family Accipitridae (Hawks and Eagles). In: del Hoyo J, Elliott A, Sargatal J (eds) Handbook of the birds of the world. Lynx Edicions, Barcelona

Tonni EP, Noriega JI (1998) Los cóndores (Ciconiformes, Vulturidae) de la región pampeana de la Argentina durante el Cenozoico tardío: distribución, interacciones y extinciones. Ameghiniana 35:141–150

Wetmore A (1927) Fossil birds from the Oligocene of Colorado. Proc Colorado Mus Nat Hist 7:3–13

Wiens JA (1982) On size ratios and sequences in ecological communities: are there no rules? Ann Zool Fennici 19:297–308

Wilson D (1975) The adequacy of body size as niche difference. Am Nat 109:769–784

Vezina AF (1985) Empirical relationships between predator and prey size among terrestrial vertebrate predators. Oecologia 67:555–565

Vizcaíno SF, Fariña RA (1999) On the flight capabilities and distribution of the giant Miocene bird *Argentavis magnificens* (Teratornithidae). Lethaia 32:271–278

Chapter 9
Bio-Connections Between Southern Continents: What is and What is Not Possible to Conclude

Geological data documents a dynamic physical environment on all timescales from the Cretaceous through the Holocene for South America. Roughly, from the Early Paleocene to the Pleistocene the South American environments showed a climate change from warm, wet, and non-seasonal (Paleocene to Eocene) to cold and dry (Middle Eocene to Early Oligocene) to seasonal climate (Middle to Late Miocene) (Ortiz Jaureguizar and Cladera 2006; Barreda and Bellosi 2003; Barreda and Palazzesi 2007) (Fig. 2.1). Terrestrial mean annual temperature for the Early Paleogene ranged between 16 and 22 °C to southern South America, 23–27 in the north and 9–15 in Antarctica (Hawking et al. 2006). Lowland swamp and fluvial environments characterized Paleocene scenarios. An increase of uplands started in the Late Eocene (ca. 40 million years ago). These changes were apparently accompanied by a transition from balanced subtropical woodlands and grasslands to predominantly grassland ecosystems.

The geologic development over time of the Andes Mountain has modified the global and regional climates; the mountain forms a pronounced topographic barrier to the atmospheric circulation and the distribution of rainfall (Graham 2005). Complete Andean orogeny occurred in pulses; the Patagonian Andes' main uplift dates to 26–28 million years ago and the most recent event was dated to 6–10 million years ago (Blisniuk et al. 2005). The rain shadow effects of the former created a narrow area with low precipitation (<300 mm year) which extends from 2 to 52 degrees south (Chacón et al. 2012) known as the South American Arid Diagonal. Geologic data shows that this climatic pattern seems to have begun to settle from the Miocene uplift (ca. 15 million years ago). Since then, a drastic increase in aridity occurred (Blisniuk et al. 2005). If this is so, it could be expected that the biota exhibit some appearances, extinctions, extirpations, or changes in some (or all) of its components. At this point, the question is how much of the avian fossil record corresponds (or if it does) with these geographic and physiognomic patterns. We cannot fully answer this question, but we can offer some plausible hypotheses and explanations.

C. P. Tambussi and F. J. Degrange, *South American and Antarctic Continental Cenozoic Birds*,   103
SpringerBriefs in Earth System Sciences, DOI: 10.1007/978-94-007-5467-6_9,
© The Author(s) 2013

At present only Neotropics hosts more than 3,150 species of birds. As seen in the previous sections of this work, the South American Cenozoic avian fossil record is far from reaching a similar figure. The largest gap in our knowledge concerns very small birds, which usually are preserved only under special circumstances and finding depends on delicate and meticulous samples. Even remotely, we could make reasonable assumptions about the origin of the current diversity patterns based solely on the fossil record and, in coincidence with many other authors (e.g., Cracraft 2001), we can argue that it is inevitable to take into account the phylogenetic information to build one scientifically accurate idea. Hence, the evolutionary timeframe of the avian biota can only be inferred using alternative time constraints (Hunn and Upchurch 2001). But it is possible to analyze, understanding the limitations imposed by the fossil evidence, some bio-connections that may have occurred during the Cenozoic while the Andean orogeny and the continental drift modeled the South American and Antarctic scenarios.

The South American's avian fossil record is scarce (Mayr 2007, Mayr 2009; Tambussi 2011) and it is particularly to the pivotal interval of time for plant and animal evolution, the Paleogene. As such, this record sheds little light on the critical interval when the ancestral stocks of many extant clades are thought to have arrived or diversified in South America.

The earliest record of Neornithes for South America comes from sediments for the Danian-Selandiano limit ($\sim$ 61.7 Ma) (Degrange et al. 2006) and specimens consist of two downy feathers. Late Paleocene Itaboraian ($\sim$ 58 Ma) birds are the oldest South American fossil Neornithes association represented by skeletal remains. Only postcranial elements of four genera are known (*Diogenornis*, *Paleopsilopterus*, *Eutreptodactylus,* and *Itaboravis*), none of which has been reported from other Paleocene avifaunas (Mayr et al. 2011). Both *Diogenornis fragilis* and *Paleopsilopterus itaboraiensis* have no (or little) flight capabilities.

For biogeographers, dispersal strategies of the biota are critical to establishing landmasses' bio-connections. Whereas flightless birds have to depend on tectonic plate movement or land bridges to get around, flight birds can use other dispersion strategies (active flying, soaring, swimming, or rafting). Due to their long-distance dispersal capabilities, flying birds do not provide critical evidence for revealing ancient biogeographic patterns. At the opposite end are the obligate terrestrial birds. Over geologic time, the loss of flying ability has taken place in many different families of birds and members of which 18 extant bird families are flightless. Indeed, five of the South American Paleogene birds recorded were supposedly flightless.

Returning to *Diogenornis*, it is the oldest known ratite and has been identified as a rheid (Alvarenga 1983) or a ratite related to the Casuariidae (Alvarenga 2010). Living ratites include five species of kiwis (New Zealand), three species of cassowaries (Australia and New Guinea), the emu (Australia), two species of rheas (South America), and the ostrich (Africa) plus two recently extinguished groups, the New Zealand moas and the elephant birds from Madagascar. All these species are flightless and they have a current distribution on isolated southern land masses.

The prevailing vision is that ratites are monophyletic, with the flighted tinamous as their sister group, suggesting a single loss of flight in the common ancestry of ratites (but see Harshman et al. 2008).

Ornithologists have hypothesized that all ratites descended from a flightless (or poor flight) ancestor that was widespread in Gondwana (see for instance Cracraft 2001; Bourdon et al. 2009; Cubo 2003; Laurin et al. 2012), and vicariance biogeography proved to be congruent with the ages estimated by the use of fossils. Although the proposed phyletic branching patterns for ratites do not fit perfectly to the timing of movements of the landmasses during the breakup of Gondwana, the continental drift as a mechanistic explanation for ratite distribution is close to be the most convenient (and popular in the literature as example of vicariance, e.g., large-scale geological events that influenced the distributions of multiple groups simultaneously). For Harshman et al. (2008) the scenario is very different in both the polyphyly of ratites and the loss of flight in more than one lineage. In this context, living ratites are evidence of parallel evolutionary trajectories from flighted ancestors. In any case, this dispute could only be resolved by collecting more data or combining molecular and morphological data which results in a consensus phylogeny.

Briggs (2003) suggested an early tertiary origin in South America for all the ratites (because the oldest fossil record of Brazil) and a subsequent broad distribution since the Paleocene, involving birds with some flying abilities that could have reached North America and Europe. After that, they could have extended southward to Africa and Madagascar. Such an early dispersal may have been followed by a later Eocene migration from South America to Australia via Antarctica.

Southern South America and Australia remained in contact through Antarctica until the Paleogene (Woodburne and Case 1996) and dispersal was still potential until the Oligocene (30–28 Ma) when the Drake Passage opened between these continents allowing the establishment of the Antarctic circumpolar current and the onset of the first Antarctic glaciations. The Antarctic migration route is indicated by the report of an Eocene ratite from the Antarctic Peninsula (Tambussi et al. 1994). The presence of the ratite strongly supports the idea that West Antarctica was used as dispersal route for obligate terrestrial organisms (Fig. 9.1). Moreover, the trans-Antarctic dispersal between Australia and South America was significantly more frequent than any other dispersal events involving other austral landmasses (San Martin and Ronquist 2004). Paleogene fossil marsupials (Woodburne and Case 1996) further document the importance of the trans-Antarctic dispersal route.

Hawkins et al. (2005), Hawking et al. (2006) proposed that "species in clades that initially evolved in the warm, wet climates of the Late Cretaceous to Eocene have been extirpated from areas that have undergo the most severe climate change, whereas clades that arose during the Miocene have radiated in areas were climates became drier" (Hawking et al. 2006: 771). Perhaps this was the case of the Eocene continental Antarctic avifauna. The already mentioned ratites with members of Falconids Polyborinae (the only representative of the non-aquatic zoophagous

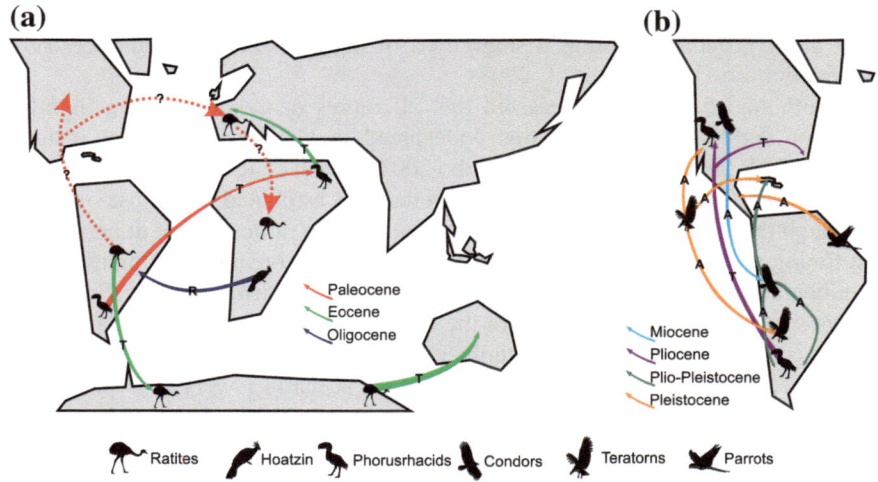

Fig. 9.1 Main avian dispersions inferred in South America during the Cenozoic according to the fossil record. **a** Paleogene. **b** Neogene. *A* aereal (flying) dispersion; *T* terrestral dispersion; *R* dispersion through rafting

style of life within this association), Ciconiiformes, Charadriiformes, Phoenicop-teriformes, Pelagornithidae, Procellariidae, and Gaviiformes were the non-penguin avian assemblages of the Eocene of La Meseta formation. We do not know what happened after the Eocene–Early Oligocene because geological record of that time is non-existent, but the truth is that all but one (Procellariidae) of these groups are not represented in Antarctica today.

Fossils of pelagornithids (bony-toothed birds) have been reported from every continent since the Paleocene to the end of the Pliocene. They were marine gliders and maybe became extirpated from Antarctica during the ocean restructuring and climatic upheavals at the Oligocene.

Extant loons (Gaviiformes, four species of the genus *Gavia*) are foot-propelled divers found in North America and northern Eurasia. Loons had a more southerly distribution than today, and their fossils have been found in California, Florida, Italy, Austria, Chile, and Antarctica. The presence of a loon in the Eocene of La Meseta Formation constitutes the more modern record of Gaviiformes in the southern hemisphere and, also extends the permanence of this Holarctic lineage since the Cretaceous to the Eocene in the southern hemisphere (Tambussi et al. 2012). Our current understanding of the dispersal process of birds is incomplete and the position of Gaviiformes within the Neornithes context in under debate. Explanations for this distributional pattern are pending.

South America and Africa started to move away since the early Cretaceous, at about 135 Ma with the opening of the South Atlantic Ocean at the latitude of Argentina. Northern South America and Africa remained connected until the mid-Late Cretaceous (110–95 Ma) and after that, Africa began drifting northeast and collided with Eurasia in the Paleocene (60 Ma). On its side, southern South

America drifted southwest into contact with Antarctica (San Martin and Ronquist 2004). The finding of a phorusrhacids in the Eocene of Africa has fundamental connotations and revitalizes the interest to understand the evolution and distribution of these birds.

Phorusrhacids consist of several terrestrial species of Cariamiformes (an ancient order that is now known not to be closely related to any other living birds) recorded since the Eocene to the Pleistocene of Uruguay, Argentina, Brazil, United State of America, and Africa. Previous reports of European phorusrhacids (Mourer Chauviré 1981; Peters 1987) have been dismissed but now, a re-examination of a fragmentary tarsometatarsus and several phalanges of old collections allow Angst and Buffetaut (2012a, b) to support again the presence of phorusrhacids during the Middle Eocene of France. If this is so, this record strongly suggests that the European bird was a trans-Tethyan Paleogene scattered. Faunal exchange between Africa and Europe during the Eocene is also sustained by other evidences (turtles, ziphodont crocodiles, placental, and marsupial mammals, see Gheerbrant and Rage 2006). Therefore, it is not surprising that an animal of the *Patagornis* size traversed the distance between Africa and Laurasia at that time and was one of the many African migrants to Europe. The position of this European taxon in the context of Phorusrhacidae is still unknown. And so is it of the newly discovered African phorusrhacid, *Lavocatavis africana* named on the basis of a femur (Mourer-Chauviré et al. 2011) recovered in Late Eocene sediments of the Glib Zegdou formation in western Algeria ($\sim$48 Ma).

But the identification of the center of origin of the family Phorusrhacidae remains highly hypothetical, the fossil evidence might suggest a South American origin (?Early Paleocene). While only two species are known from the early Eocene of South America, all types of Phorusrhacids are present at the Oligocene, even those which seem more derived (Degrange 2012). Then it could be assumed a rapid Eocene radiation which worsened during Oligocene times. Assuming that the systematic position of *Lavocatavis* is correct, it would imply a very early dispersal from America to Africa. This is one of the hypotheses of Mourer-Chauviré and coauthors (2011) and involves scattering mechanisms.

South America and Africa were already separated by the time that *Lavocatavis* evolved (Eocene). The last link that connected Africa to another continent (i.e., South America) was by Cretaceous (Albian–Aptian) ruling out a land migration from America to Africa. Throughout 75 million years, between the Mid-Cretaceous to the Early Miocene, Africa was isolated (Gheerbrant and Rage 2006). One possibility is that the precursor of *Lavocatavis* arrived in Africa by rafting across the prehistoric Atlantic Ocean from east to west. Although some authors dismiss the possibility of survival in open ocean waters because solar radiation, intense changes in temperature, or lack of food or water, it is conceivable that terrestrial vertebrates on floating islands survive long oceanic transport (Kappeler 2000; Samonds et al. 2012).

Prior to 48 million years ago, the northeast portion of South America was significantly closer to Africa (ca. 1,000 km) than it is at present (2,600 km) and very large islands that existed on what are now the submerged Rio Grande Rise and Walvis Ridge (de Oliveira et al. 2008). These Islands may have provided a stepping-stone route across the proto-Atlantic Sea to the hypothetical *Lavocatavis* flightless ancestor.

Another option is that the ancestors of the African form could fly [in the same way as some of the South American Psilopterinae could do it according to Tonni and Tambussi (1988) and Degrange (2012)], and perhaps may have flown to Africa and evolved in more or less the same way as those which remained in South America. This would imply that two separate lineages of the same group evolved in the same manner (i.e., tendency to flightless and cursoriality). Although a cursorial lifestyle can result in morphological convergence or parallelism, morphology of the femora of *Lavocatavis* and that of the remainder phorusrhacids are almost exactly the same. Multiple losses of flying capabilities, with the implication of greater dispersal aptitudes for ancestral phorusrhacids, make this model less compelling.

We are thinking in a third potential scenario following others' reasoning. First, Africa appears to have been the first continent that became isolated in the breakup of Gondwana and, at least from a paleobiological point of view, it was more strongly related to Laurasia than to any Gondwanan continent (Gheerbrant and Rage 2006). Second, it has long been recognized that adaptation to a flightless, cursorial lifestyle can result in morphological convergence or parallelism, especially in the postcranial skeleton (Davies 2002). Taking these altogether, a possible alternative scenario could be considered in which the European and African forms were not related with the American Phorusrhacidae, and the similarity could reflect convergent adaptation to a flightless, cursorial lifestyle (such us those hypothesized in some European non-phorusrhacid Cariamiformes as *Salmila* or *Elaphrocnemus*). Unfortunately we do not have good phylogenies containing all the extinguished Cariamiformes (and other giant and flightless birds) and the expectations are low because little more than a nearly-complete right femur is available (in the case of *Lavocatasis*) which is not a high informative bone in the systematic field. Although all these hypotheses have been built by making use of pertinent arguments, this does not allow a single simple explanation for the connections between the South American and African forms, and the problem still remains.

Aves from the richly fossiliferous Santa Cruz Formation (Late-early Miocene), are taxonomically the most diverse in terms of known avian genera of all Tertiary faunas on South America. Nine families, 15 genera and at least 18 species were present at the Santacrucian association. Four of them correspond to phorusrhacids and the remainder 13 belong to ecomorphs which are also represented in modern faunas (Tambussi 2011). Many different data sources (carbon isotopes, Blisniuk et al. 2005; fossil xenarthrans, Vizcaíno et al. 2006; palynological information, Barreda and Palazzesi 2007) allowed to characterize the Santacrucian environments with a balanced presence of woodlands and grasslands in relatively dry conditions.

Representatives of some of these families have current distributions further north. To name some, darters (Anhingidae) have a Pantropical distribution with only two species (*Anhinga anhinga* and *A. melanogaster*) living in tropical

America, Africa, Asia, and Australia. The distribution of the sister taxon of the extinct *Thegornis musculosus* (Falconidae), the Laughing Falcon *Herpetotheres*, is found from both coastal Mexico through Central and South America to Paraguay and northern Argentina. The modern representative of *Cariama* (*Cariama cristata*) is found from eastern Brazil to central Argentina. Again, this seems to reflect the assumption of Hawkins and colleagues on the diversification of fauna during the Miocene: "...clades that arose during the Miocene have radiated in areas were climates became drier" (Hawking et al. 2006: 771).

The early appearance of a typically Pampean bird fauna recognizable at the Pampean region is estimated at 10 Ma to ~6 million years at approximately 35° S. Possible scenarios for that time are made up of open environments, maybe xerophyllous shrubby steppes and some forest (Cenizo and Agnolín 2010).

North and South America became connected during the middle Cretaceous across the proto-Caribbean archipelago, located to the west than it is at present. The connection was interrupted by the early Eocene about 49 million years ago. Significant faunal exchange would have taken place across both Americas involving a first dispersal of placental and marsupial mammals. About 15 million years ago, both Americas became connected again via the Panama Island Arc (a series of emergent platforms that made up transitory, discontinuous land routes) and then across the Panama Isthmus (Montes et al. 2012). This new connection established a new biotic exchange known as the Great American Interchange (GAI) which is thought to have been predominantly southward (but read below).

The deep effect of the GAI in shaping New World mammalian diversity is well documented due to the fossil record, but its influence on the avian diversity patterns of South America is still a matter of opinion since the classic work "Fossil and recent avifaunas and the interamerican exchange" written over three decades ago by Vuilleumier (1985).

As we mentioned, birds lack a good fossil record, especially from tropical latitudes, and consequently the effects of land bridge formation on avian interchange is difficult to understand. Because of its ability to fly, birds are good dispersers and it was thought that the isolation between the Americas and its subsequent connection, would not have influenced the distribution of birds. The discussion of the importance of this event in exchange of birds remained in a great impasse. Recent advances in the field of phylogeography (the history and formation of species) aroused new hopes although results are emerging.

Few recent avian molecular studies focused on the relevance of the formation of the isthmus in avian diversification patterns but are restricted to few families of passerine (e.g., Barker 2007; Burns and Racicot 2009; Weir et al. 2008) or trogons (DaCosta and Klicka 2008). The main focus of those contributions was to evaluate the timing of diversification and direction of dispersals and they arrived at non-linear results. For example, hypotheses include that most dispersal events occurred before the final isthmus formation (Weir et al. 2008), after (Barker 2007; DaCosta and Klicka 2008), or during final closure (Pérez-Emán 2005). Evaluations about rates of interchange of Weir and colleagues (2009) show an increment after land bridge completion in tropical forest-specializing groups, but not in habitat generalists. Also, they proposed a south to north direction of the traffic after the isthmus completion (in contrast to mammals) but a north to south transfer before. The later study is restricted to four families of passerines (although it is not stated in the title) and makes clear that the history of the birds across the Panamanian Isthmus must have been complex and different for each individual group. As always, generalizations could be controversial.

A different source of molecular evidences provides contradictory explanations about early passerines diversification. Ericson et al. (2002) recognized both east and west Gondwanan suboscines clades, being the neotropical suboscines (Tyrannides) descendants of the latter. Irestedt et al. (2002: 507) said: "the South American groups could well have originated in the far south of that continent, or even on Antarctica, which had a pleasant climate up until the early Miocene". This view serves as a fuel injection to paleontologists who look for fossils every austral summer in Antarctica.

Crisci et al. (1991) suggested that South American biota have two components with different biogeographic affinities, one northern tropical and a second southern temperate component. The dispersal events between these two components are lower than that between northern South America and the Holarctic (San Martin and Ronquist 2004). Again, partial responses returned from fossils and it is important to recognize the complexity of the situation rather than a priori assume a single explanation.

Condors seem to have had a North American origin with a posterior shift to South America possibly using the coastal western side of the Andes (Emslie 1988; Stucchi and Emslie 2005; Tambussi 2011). The fossil record suggested that they were more diverse in the past and *Perugyps diazi* indicates that condors were present in South America by the Late Miocene to early Pliocene. During the Plio-Pleistocene they speedily spread out of the continent, occupying much of the Pampas and Brazil where their record fossil is abundant (Tambussi and Noriega 1999; Tonni and Noriega 1998). The interpretation of the fossil evidence suggests dispersion prior to the permanent establishment of the isthmus but dispersal may have been more important later.

Galliformes fossil record is near absent in South American localities even of the two most primitive families with predominantly southern hemisphere distribution (Megapodiidae and Cracidae). Interestingly, the fossil record of galliformes is significantly older in the northern hemisphere than it is in Gondwana (Mayr 2009).

Fossil parrots are very scarce and restricted to the Pleistocene of the Pampean Region in Argentina and Uruguay. Molecular studies were also applied to the Neotropical *Amazona* (Psittaciformes), and divergence-date estimates that diversification of the group in South America occurred rapidly during the Pleistocene but they arrived in Middle America after the connection between both landmasses (Eberhard and Bermingham 2004).

Most extant South American bird lineages are not known from the fossil record. Phylogenies that include South American birds are underrepresented in the available molecular studies. Our current comprehension of dispersal processes is still fragmentary and we need to understand the underlying mechanisms in much more detail. The relative significance of dispersal, vicariance, extirpations, or extinctions in shaping the evolutionary history and/or biogeographic pattern of birds remains hard to evaluate. And in many aspects, the fossil record remains silent.

References

Alvarenga HMF (1983) Uma ave ratite do Paleoceno brasileiro: bacia calcária de Itaboraí, estado do Rio de Janeiro, Brasil. Bol Mus Nac do Rio de Jan, Geol 41:1–47

Alvarenga HMF (2010) Diogenornis fragilis Alvarenga, 1985, restudied: a South American ratite closely related to Casuariidae. In: 25th Int Ornit Cong, p 143

Angst D, Buffetaut E (2012a) A large Phorusrhacid bird from the middle Eocene of France. In: Worthy TH, Göhlich UB (eds) Abstracts of the 8th international meeting of the society of the Avian Paleontology and evolution

Angst D, Buffetaut E (2012b) A large phorusrhacid from the middle Eocene of France and its palaeobiogeographical implications. 26 Jor Arg Pal Vert, Abstracts available in CD

Barker FK (2007) Avifaunal interchange across the Panamanian isthmus: insights from Campylorhynchus wrens. Biol J Linn Soc Lond 9:687–702

Barreda V, Bellosi E (2003) Ecosistemas terrestres del Mioceno temprano de la Patagonia central: primeros avances. Rev Mus Arg Cienc Nat 5:125–134

Barreda V, Palazzesi L (2007) Patagonian vegetation turnovers during the Paleogene-Early Neogene: origin of arid adapted floras. Bot Rev 73:31–50

Blisniuk PM, Ster LA, Chamberlain CP, Idleman B, Zeitler PK. (2005) Climatic and ecologic changes during Miocene surface uplift in the southern Patagonian Andes. Earth Planet Sci lett 230:125–142

Bourdon E, de Ricqlès A, Cubo J (2009) A new transantarctic relationship: morphological evidence for a Rheidae–Dromaiidae–Casuariidae clade (Aves, Palaeognathae, Ratitae). Zool J Linn Soc 156:641–663

Briggs JC (2003) Fishes and birds: Gondwana life rafts reconsidered. Syst Biol 52:548–553

Burns K, Racicot R (2009) Molecular phylogenetics of a clade of lowland tanagers: implications of the avian participations in the Great American Interchange. Auk 126:635–648

Cenizo MM, Agnolín FL (2010) The southernmost records of Anhingidae and a new basal species of Anatidae (Aves) from the lower–middle Miocene of Patagonia, Argentina. Alcheringa 34:1–22

Chacón J, Camargo de Assis M, Meerow A, Wand Renner SS (2012) From east Gondwana to central America: historical biogeography of the Alstroemeriaceae. J Biogeogr. doi:10.1111/j.1365-2699.2012.02749.x

Cracraft J (2001) Avian evolution, Gondwana biogeography and the Cretaceous-tertiary mass extinction event. Proc Royal Soc London 268:459–469

Crisci J, Cigliano MM, Morrone J, Roig S (1991) Historial biogeography of southern south America. Syst Zool 40:152–171

Cubo J (2003) Evidence for speciational change in the evolution of ratites (Aves: Palaeognathae). Biol J Linn Soc 80:99–106

Dacosta J, Klicka J (2008) The Great American Interchange in birds: a phylogenetic perspective with the genus Trogon. Mol Ecol 17:1328–1343

Davies SJJF (2002) Ratites and tinamous: Tinamidae, Rheidae, Dromaiidae, Casuariidae, Apterygidae, Struthionidae. Oxford University Press, Oxford

de Oliveira FB, Molina EC, Marroig G (2008) Paleogeography of the South Atlantic: a route for primates and rodents into the new world? In: Garber PA, Estrada A, Strier KB (eds) South American primates: comparative perspectives in the study of behavior, ecology, and conservation. Springer, New York

Degrange FJ (2012) Morfología del cráneo y complejo apendicular posterior de aves fororracoideas: implicancias en la dieta y modo de vida. Dissertation, Universidad Nacional de La Plata

Degrange F, Tambussi C, Iglesias A, Zamuner A, Wilf P (2006) Primer registro de Aves para el Daniano. Ameghiniana 43:13R

Eberhard J, Bermingham E (2004) Phylogeny and biogeography of the amazona ochrocephala (Aves: Psittacidae) complex. Auk 121:318–332

Emslie S (1988) A early condor-like vulture from North America. Auk 105:529–535

Ericson PGP, Cooper A, Christidis L, Irestedt M, Jackson J, Johansson US, Norman J (2002) A Gondwanan origin of passerine birds supported by DNA sequences of the endemic New Zealand wrens. Proc R Soc Lond B 264:235–241

Gheerbrant E, Rage J (2006) Paleobiogeography of Africa: how distinct from Gondwana and Laurasia? Palaeogeog Palaeoecol 24:224–246

Graham A (2005) The Andes: a geological overview from a biological perspective. Ann Miss Bot Gard 96:371–385

Haq BU, Hardenbol J, Vail PR (1987) Chronology of fluctuating sea levels since the Triassic (250 million years ago to present). Science 235:1156–1167

Harshman J, Braun EL, Braun MJ, Huddleston CJ, Bowie RCK, Chojnowski JL, Hackett SJ, Han K, Timball RT, Marks BD, Miglia KJ, Moore WS, Reddy S, Sheldon FH, Steadman DW, Steppan SJ, Witti CC, Yuri T (2008) Phylogenomic evidence for multiple losses of flight in ratite birds. PNAS 105:13462–13467

Hawking B, Diniz-Flho FA, Jaramillo C, Soeller S (2006) Post-Eocene climate change, niche conservatism, and the latitudinal diversity gradient to new world birds. J Biogeog 33: 770–780

Hawkins BA, Diniz-Filho JAF, Soeller SA (2005) Water links the historical and contemporary components of the Australian bird diversity gradient. J Biogeogr 32:1035–1042

Hunn C, Upchurch P (2001) The Importance of time/space in diagnosing the causality of phylogenetic events: towards a "chronobiogeographical" paradigm? Syst Biol 50:391–407

Irestedt M, Fjelds J, Johansson UF, Ericson PGP (2002) Systematic relationships and biogeography of the tracheophone suboscines (Aves: Passeriformes). Mol Phylog Evol 23:499–512

Kappeler PM (2000) Lemur origins: rafting by groups of hibernators? Folia Primatol 71:422–425

Laurin M, Gussekloo SWS, Marjanović D, Legendre L, Cubo J (2012) Testing gradual and speciational models of evolution in extant taxa: the example of ratites. J Evol Biol 25:293–303

Mayr G (2007) The birds from the Paleocene fissure filling of Walbeck (Germany). J Vert Paleontol 27:394–408

Mayr G (2009) Paleogene Fossil Birds. Springer-Verlag, Berlin Heidelberg

Mayr G, Alvarenga HMF, Clarke J (2011) An Elaphrocnemus-like landbird and other avian remains from the late Paleocene of Brazil. Acta Palaeontol Pol 56:679–684

Montes C, Bayona G, Cardona A, Buchs DM, Silva CA, Morón S, Hoyos N, Ramírez DA, Jaramillo CA, Valencia V (2012) Arc-continent collision and orocline formation: closing of the central American seaway. J Geophys Res. doi:10.1029/2011JB008959

Mourer Chauviré C (1981) Premire indication de la présence de Phorusrhacides, famille d'oiseaux géants d'Amerique du Sul, dans le Tertiaire européen: Ameghinornis nov.ge (Aves, Ralliformes) des Phosphorites du Quercy. Fr Geobios 14:637–647

Mourer-Chauviré C, Tabuce R, Mahboubi M, Adaci M, Bensalah M (2011) A Phororhacoid bird from the Eocene of Africa. Naturwissenschaften. doi:10.1007/s00114-011-0829-5

Ortiz Jaureguizar E, Cladera GA (2006) Paleoenvironmental evolution of suthern South America during the Cenozoic. J Arid Environ 66:498–532

Pascual R, Carlini AA, Bond M, goin FJ (2002) Mamíferos cenozoicos. In: Haller MJ (ed) Geologia y Recursos Naturales de Santa Cruz. Asociación Geológica Argentina, Buenos Aires

Pérez-Emán JL (2005) Molecular phylogenetics and biogeography of the neotropical redstarts (Myioborus; Aves, Parulinae). Mol Phylog Evol 37:511–528

Peters DS (1987) Ein "Phorusrhacidae" aus dem Mittel-Eozan von Messel (Aves, Gruiformes, Cariamae). Doc Lab Géol Lyon 99:71–87

Samonds K, Godfreyb L, Alic J, Goodmand S, Vencesf M, Sutherlandb M, Irwing M, Krause D (2012) Spatial and temporal arrival patterns of Madagascar's vertebrate fauna explained by distance, ocean currents, and ancestor type. PNAS. doi:10.1073/pnas.1113993109

San Martin I, Ronquist F (2004) Southern hemisphere biogeography inferred by event-based models: plant vesus animal patterns. Syst Biol 53:216–243

Stucchi M, Emslie SD (2005) Un Nuevo Cóndor (Ciconiiformes, Vulturidae) del Mioceno Tardío-Plioceno temprano de la Formación Pisco, Perú. The Condor 107:107–113

Tambussi CP (2011) Paleoenvironmental and faunal inferences based upon the avian fossil record of Patagonia and Pampa: what works and what does not. Biol J Linn Soc 103:458–474

Tambussi CP, Noriega JI (1999) The fossil record of condors (Aves, Vulturidae) of Argentina. Smith Cont Pal 89:171–184

Tambussi CP, Noriega JI, Gazdzicki A, Tatur A, Reguero MA, Vizcaíno SF (1994) Ratite bird from the Paleogene La Meseta formation, Seymour Island, Antarctica. Pol Polar Res 15:15–20

Tambussi CP, Degrange FJ, Reguero MA, Marenssi SA, Santillana SN (2012) Antarctic Eocene loon (Gaviiformes): last refuge of survivor of a long typically Holarctic lineage? SCAR open science conference (OSC), Portand. http://scar2012.geol.pdx.edu/themes.php. Accessed 1 May 2012

Tonni EP, Noriega JI (1998) Los cóndores (Ciconiformes, Vulturidae) de la región pampeana de la Argentina durante el Cenozoico tardío: distribución, interacciones y extinciones. Ameghiniana 35:141–150

Tonni EP, Tambussi CP (1988) Un nuevo Psilopterinae (Aves: Ralliformes) del Mioceno tardío de la provincia de Buenos Aires, República Argentina. Ameghiniana 25:155–160

Vizcaíno SF, Bargo MS, Kay RF, Milne N (2006) The armadillos (Mammalia, Xenarthra, Dasypodidae) of the Santa Cruz Formation (early-middle Miocene): an approach to their paleobiology. Palaeogeog Palaeoecol 237:255–269

Vuilleumier F (1985) Fossil and recent avifaunas and the inter American exchange. In: Stehli FG, Webb SD (eds) The great American biotic interchange. Plenum, New York

Weir JT, Bermingham E, Miller MJ, Klicka J, González MA (2008) Phylogeography of a morphologically diverse neotropical montane species, the common bush-tanager (Chlorospingus ophthalmicus). Mol Phylog Evol 47:650–664

Weir J, Berminghamb E, Schluter D (2009) The great American biotic interchange in birds. PNAS 106:21737–21742

Woodburne MO, Case JA (1996) Dispersal, vicariance, and the late Cretaceous to early tertiary land mammal biogeography from South America to Australia. J Mamm Evol 3:121–161

Zachos J, Shackleton NJ, Revenaugh JS, Pälike H, Flower BP (2001) Climate response to orbital forcing across the Oligocene-Miocene. Science 292:274–278